TECHNICAL STATICS AND STRENGTH OF MATERIALS
Second Edition

James R. Thrower, Jr.
Massasoit Community College

PWS PUBLISHERS

Prindle, Weber & Schmidt • Duxbury Press • PWS Engineering • Breton Publishers
Statler Office Building • 20 Park Plaza • Boston, Massachusetts 02116

Copyright © 1986 PWS Publishers, Boston, Massachusetts 02116. Copyright © 1976 Wadsworth, Inc. All rights reserved. No part of this book may be reproduced, stored in a retrieval system, or transcribed, in any form or by any means—electronic, mechanical, photocopying, recording, or otherwise—without the prior written permission of PWS Publishers.

PWS Publishers is a division of Wadsworth, Inc.

Library of Congress Cataloging-in-Publication Data

Thrower, James R.
   Technical statics and strength of materials.

   Includes index.
   1. Statics. 2. Strength of materials. I. Title.
TA351.T48   1986   620.1'03   85-31352
ISBN 0-534-06384-5

Printed in the United States of America
1 2 3 4 5 6 7 8 9—90 89 88 87 86

ISBN 0-534-06384-5

Acknowledgments

Chapter 4: Photo 4.1, courtesy of Bethlehem Steel Corporation. **Chapter 7:** Figures 7.1 and 7.3, courtesy of Tinius Olsen Testing Machine Company, Inc.; Figures 7.4 and 7.5, from Laurson and Cox, *Mechanics of Materials,* copyright © 1962 by John Wiley & Sons, reprinted by permission of John Wiley & Sons, Inc.; Figure 7.8, from *Design of Machine Elements* by M.F. Spotts, © 1961 by Prentice-Hall, Inc.; Figures 7.9 and 7.10, from *Design of Machine Members* by A. Vallance, © 1938 by McGraw-Hill Book Company. **Chapter 8:** Figure 8.2, from Baumeister and Marks, *Standard Handbook for Mechanical Engineers,* © 1967 by McGraw-Hill Book Company. **Chapter 10:** Photo 10.2, courtesy of Boston Gas Company. **Chapter 12:** Photo 12.3, courtesy of General Electric Company. **Chapter 15:** Photo 15.4, courtesy of Vishay Research and Education. **Chapter 19:** Table 19.1, from Baumeister and Marks, *Standard Handbook for Mechanical Engineers,* 8th edition, © 1978 by McGraw-Hill Book Company; Figure 19.3, from *Elements of Strength of Materials,* Timoshenko and Young, © 1962, reprinted by permission of D. Van Nostrand Company; Photo 19.5, courtesy of Bethlehem Steel Corporation. **Chapter 20:** Figure 20.9, graph from *Mechanics of Materials,* E. P. Popov, © 1952 by Prentice-Hall, Inc.

Sponsoring editor: George J. Horesta; Production supervision: Technical Texts, Inc.; Production editor: Jean T. Peck; Text design: Mary S. Mowrey; Cover photo: Morton Beebe/The Image Bank; Composition: Crane Typesetting Service, Inc.; Cover printing: New England Book Components, Inc.; Text printing and binding: Halliday Lithograph.

This text is dedicated to my **parents**.

Contents

PREFACE   xiii

1 STARTING OUT   1
    Objectives   1
    Introduction   1
    Suggestions for Mathematical Operations   3
        Procedure for Solving Problems   7
    Metric Conversion   8

2 DEFINITIONS, EQUILIBRIUM, FREE BODY DIAGRAMS, AND CONCURRENT FORCE SYSTEMS   9
    Objectives   9
    Introduction   9
    Definitions   10
    Free Body Diagrams   14
    Mechanical Equilibrium   17
    Concurrent Force Systems   18
        Procedure for Force Polygon Method   19
        Procedure for Component Summation Method   22
    Summary   27
    Problems   27
    Computer Program   32

3 NONCONCURRENT FORCE SYSTEMS   33
    Objectives   33
    Introduction   33
    Moment of a Force   33
    Parallel Force Systems   36
        Procedure for Finding Reactions of Beams on Two Supports   38
    Pulleys   40
    Nonconcurrent Force Systems   43
        Procedure for Solving Nonconcurrent Force Problems   43
    Replacing One Force System with Another   45
        Procedure for Replacing a Nonconcurrent Force System   47
    Summary   49

vi   CONTENTS

    Problems   50
    Case Study   57
    Computer Program   59

4 FRAMED STRUCTURES AND FRICTION   60
    Objectives   60
    Introduction   60
    Two-Force Members   61
    Truss Structures   62
        Procedure for the Method of Joints   66
    Zero-Force Members   66
    Other Framed Structures   73
        Procedure for Solving Frame Problems   73
    Static Friction   76
        Procedure for Solving Friction Problems   79
    Wedges   84
        Procedure for Solving Wedge Problems   87
    Summary   89
    Problems   89
    Case Study   98
    Computer Program   98

5 CENTROIDS AND MOMENTS OF INERTIA OF AREAS   100
    Objectives   100
    Introduction   100
    Centroid—First Moment of Area   100
        Procedure for Solving Centroid Problems   104
    Moment of Inertia of an Area—Second Moment of
        Area   106
        Procedure for Finding Moment of Inertia or Radius of
            Gyration of an Area   108
    Summary   109
    Problems   110

6 STRESS, STRAIN, AND DEFORMATION   113
    Objectives   113
    Introduction   113
    Tensile, Compressive, and Shear Stresses   114
    Deformation   118
        Procedure for Solving Stress/Strain Problems   121
    Summary   123
    Problems   123
    Case Study/Computer Program   127

7 MECHANICAL PROPERTIES OF MATERIALS   129

    Objectives   129
    Introduction   129
    Static Tensile Test   130
    Carbon Steels   133
    Proportional Limit   134
    Modulus of Elasticity   135
    Elastic Limit   135
    Yield Point   135
    Ultimate Tensile Strength   136
    Necking Range   136
    Plastic Range   136
    Ductility   137
    Malleability   137
    Brittleness   137
    Stiffness   137
    Yield Strength   138
    Hardness   138
    True Stress   138
    Modulus of Resilience   139
    Modulus of Toughness   139
    Coldworking   140
    Fatigue   141
    Creep   142
    Low-Temperature Impact   144
    Poisson's Ratio   144
    Summary   146
    Problems   146
    Computer Program   147

8 ALLOWABLE STRESS, CONCENTRATED STRESS, AND STRESS ON OTHER AREAS   149

    Objectives   149
    Introduction   149
    Steel Beam Designations   150
    Allowable Stress   150
    Concentrated Stress   152
    Stress Induced on Other Areas Due to a Two-Force
        System   155
        Procedure for Sketching Mohr's Circle   157
    Summary   160
    Problems   161
    Computer Program   164

viii   CONTENTS

9 STATICALLY INDETERMINATE MACHINE AND STRUCTURAL MEMBERS   165

    Objectives   165
    Introduction   165
    Statically Indeterminate Machine Members   165
       Procedure for Solving Statically Indeterminate Problems   167
    Effects of Temperature Change   172
       Procedure for Solving Temperature Problems   173
    Summary   176
    Problems   177
    Case Study   182
    Computer Program   183

10 THIN-WALLED PRESSURE VESSELS   184

    Objectives   184
    Introduction   184
    Bolted Covers of Pressure Vessels   184
    Tensile Stress on a Circumferential Seam or Joint   186
       Procedure for Deriving the Stress Formula for a Circumferential Seam   186
    Tensile Stress on a Longitudinal Seam or Joint   190
    Summary   193
    Problems   193
    Computer Program   196

11 WELDED AND RIVETED JOINTS   198

    Objectives   198
    Introduction   198
    Welded Joints   198
       Procedure for Solving Welded Joint Problems   202
    Riveted Joints   204
       Procedure for Solving Riveted Lap Joint Problems   207
       Procedure for Solving Riveted Butt Joint Problems   209
    Problems   213
    Computer Program   217

12 TORSION ON CIRCULAR SHAFTS AND COUPLINGS   218

    Objectives   218
    Introduction   218
    New Symbols   218

Shear Stress on a Shaft Cross Section 219
Torsion Formula 1, Relating Shear Stress to Torque 221
Torsion Formula 2, Relating Angle of Twist to Stress 225
Couplings 228
Longitudinal Shear Force 230
Summary 232
Problems 233
Computer Program 236

13 SHEAR AND MOMENT DIAGRAMS 238
Objectives 238
Introduction 238
Shear Diagrams 241
 Procedure for Constructing Shear Diagrams 241
Moment Diagrams 246
 Procedure for Constructing Moment Diagrams 246
Relationships Between Shear and Moment Diagrams 247
 Procedure for Writing Moment Diagram Equations 248
Point of Inflection 252
Hinged Beam 253
Summary 255
Problems 255
Computer Program 259

14 BENDING STRESS FORMULA 261
Objectives 261
Introduction 261
Effects of Bending Stresses 261
The Bending Stress Formula 263
Summary 271
Problems 271
Computer Program 274

15 SHEAR STRESS IN A BEAM 276
Objectives 276
Introduction 276
The Beam Shear Stress Formula 278
Manufactured Steel and Aluminum Beams 284
Shear Stress in Bolts in a Built-Up Beam 285
Limitations 288
Summary 289

Problems 289
Computer Program 292

16 BEAM DESIGN 293
Objectives 293
Introduction 293
Use of Section Modulus in Design 294
   Design Procedure 294
Summary 298
Problems 298
Computer Program 300

17 BEAM CURVATURE AND DEFLECTION FORMULAS 302
Objectives 302
Beam Curvature 302
Beam Deflection 306
   Procedure for Deriving the First Moment-Area Formula 308
   Procedure for Deriving the Second Moment-Area Formula 309
   Procedure for Solving Beam Deflection Problems 311
Summary 320
Problems 321
Case Study 324
Computer Program 325

18 STATICALLY INDETERMINATE BEAMS 326
Objectives 326
Introduction 326
Beam Fixed at One End and Supported at the Other 328
   Procedure for Solving for Reactions at the Supported End of a Beam 328
Moment Diagram by Parts 331
Two-Span Continuous Beam 334
   Procedure for Solving for Reactions of a Two-Span Continuous Beam 335
Beam Fixed at Both Ends 339
Complete Shear and Moment Diagrams 343
Summary 344
Problems 345
Computer Program 347

19 COLUMNS WITH CONCENTRIC LOADS   348
    Objectives   348
    Introduction   348
    Definitions   349
    Long Columns   350
    Short Columns   354
        Procedure for Solving Column Problems   357
    Compression Blocks   362
    Summary   362
    Problems   364
    Computer Program   365

20 MISCELLANEOUS TOPICS   366
    MOHR'S CIRCLE FOR STRESS DETERMINATION
        (FORCES APPLIED TO MUTUALLY
        PERPENDICULAR SURFACES)   366
    Objectives   366
    Mohr's Circle   366
        Procedure for Construction and Use of Mohr's Circle   368
    ECCENTRIC LOADS ON MACHINE MEMBERS, HELICAL
        SPRINGS, AND RIVETED JOINTS   374
    Objectives   374
    Compression Blocks and Offset and Curved Machine
        Members   374
        Procedure for Solution of Eccentric Loads on
            Compression Blocks   377
        Procedure for Solution of Loads Applied to Offset and
            Curved Members   380
    Shear Stresses in Round Wire Helical Springs   383
        Procedure for Solution of Loaded Spring Problems   386
    Eccentrically Loaded Riveted Joints   387
        Procedure for Solving Eccentrically Loaded Riveted
            Joint Problems   390
    Summary   395
    Problems   396
    Computer Program   402

APPENDIX: HANDBOOK DATA   **404**

ANSWERS TO ODD-NUMBERED PROBLEMS   **420**

INDEX   **429**

Preface

This text was written to address the need for an introductory text that combines statics and strength of materials in one volume. It is intended for use in departments of mechanical, civil, architectural, and industrial (manufacturing) technology. The presentation emphasizes descriptive coverage and problem solving. Excessive levels of theory and mathematical derivation are avoided. A working knowledge of college algebra and trigonometry is assumed of the student. No calculus is employed in the text.

Sufficient coverage of statics is introduced early enough in the presentation to facilitate student success in mastering the concepts and problems of strength of materials. Both the U.S. Customary (English) and SI Metric systems of units are employed in the text. A fairly lengthy appendix contains useful and up-to-date handbook data as well as tables of metric units.

The introductory chapter includes a review of basic mathematical topics that are a frequent source of student errors. It also covers relevant trigonometric functions and SI metric units. The remaining chapters are organized in a manner flexible enough to permit the use of this text in courses of different lengths: one quarter, one semester, or possibly two quarters. A solutions manual for the text is available from the publisher to instructors who have adopted the text for classroom use.

Several features of this text particularly suit it to the needs of engineering and industrial technology students:

1. Calculations are accompanied by both descriptive and mathematical explanations.
2. A given concept is introduced and fully developed before another new concept is introduced.
3. While the presentation does not require the derivation of formulas, the concepts used for such derivation are explained to enhance student understanding.
4. The use of fundamental, rather than specific, formulas is encouraged to avoid a "mechanistic" approach to problem solving and to deepen student awareness of concepts. In some cases, alternative methods of solution are presented.
5. Common areas of confusion (such as the distinction between brittleness and stiffness) receive extra explanation.

6. The development of concepts in the text as well as the worked examples and end-of-chapter exercise problems are accompanied by many clear line illustrations for easy visualization.

At the suggestion of many instructors and students who used the first edition, the following changes and additions have been included in this revised second edition:

1. Step-by-step procedures for problem solution have been emphasized, and the sample problems following these sections clearly indicate these steps as the solution progresses.
2. The number of abstract problems has been greatly reduced, and emphasis has been placed on practical applications.
3. Computer programs (in BASIC) have been added to most of the chapters. Once a student has mastered the concepts, these programs provide the means to tackle more involved and realistic problems.
4. Case studies have been added to several chapters. Several of these "failure" cases actually occurred, resulting in loss of life and property damage.

It is intended that these changes, particularly the reorganization of the procedures and sample problems, will greatly aid the student in mastering the subject of statics and strength of materials.

The author wishes to thank Professor Harry Haff of Massasoit Community College for his assistance and suggestions. Many other individuals have contributed their efforts to this work: Ed Francis and George Horesta of Breton Publishers have been a source of motivation and support; Nils Buus, Ed Paul, Carl Webster, and Eino Fagerlund reviewed the original manuscript and provided suggestions that greatly improved the work; the staff at Technical Texts, Inc., particularly Jean Peck, spared no efforts in the production of this revision. The author gratefully acknowledges their help.

Starting Out

OBJECTIVES

This book will introduce you to the fundamentals of statics and strength of materials and to the use of handbooks. Using handbooks is an important part of a technician's job. You will become familiar with the advantages of handbooks through the application of information from the handbook data tables included in the Appendix. You will be expected to apply your knowledge of algebra and trigonometry and your growing knowledge of the subject to identify the problem, choose the correct formulas, substitute the proper quantities, and arrive at an answer. You will not be expected to derive formulas. The theories behind the formulas are explained to deepen understanding and to increase your confidence and ability to choose and use formulas.

INTRODUCTION

In ancient times, the term *engineer* was not used. The earliest human activities now considered to involve engineering were done for military purposes—roads, bridges, canals, fortifications, and construction of military machines like the ballista, catapult, and battering ram. The separation of military and civil engineering was a gradual process. As the industrial revolution that spanned parts of the 18th and 19th centuries produced cheaper goods and cheaper and faster methods of transportation, other branches of engineering began to develop, such as mechanical, electrical, chemical, and marine engineering.

Knowledge from the fields of statics and strength of materials is used in almost all branches of engineering and can be thought of as the application of physical principles to the design of safe and economical structures. *Statics* is that division of physics dealing with forces acting on an object at rest. *Strength of materials* covers that body of engineering knowledge dealing with the internal forces and changes in dimension of engineering materials when subjected to loads. Both of these definitions will be expanded in later chapters.

Technicians in the civil, architectural, mechanical, chemical, and electrical disciplines can all use this information to assist engineers in the design, manufacture, construction, and maintenance of goods and machinery. For example, the same theories and formulas can be applied to a large beam in a bridge, a floor joist in a house, or the minute reed switches used in electronic circuits. Technicians must know that a machine part or component of a structure will be strong enough to perform its function satisfactorily but not strong to the point where the machine or structure is uneconomical or exceeds space or weight limitations.

This course provides an understanding of the formulas used in the design and analysis of members of machines or structures. Here, *design* means the selection of the right material, size, and shape of a member to support a load. *Analysis* means determining the forces both external and internal to a member already chosen for a job. In addition, the fundamentals of statics and strength of materials provide the basis for further work in structural design and machine design.

After technicians have gained experience in their field, they are often given complete charge of certain design and analysis problems. One technician specified the size of beams needed for turnpike overpasses. In another instance, a technician analyzed roof trusses in a large building and determined that reinforcing would be needed before the owner installed an air conditioning unit on the roof. Another technician was asked to determine the number and size of bolts needed for a cover on a pressure vessel the week after graduation.

It is hard to describe a typical problem, but basically the technician has to have, or gain, knowledge of (1) the external forces applied to the member, (2) the internal forces and deformations developed, and (3) the size and type of material needed to prevent rupturing.

Conversion to the metric system is under way in the United States, so a number of sample problems and student problems in metric units are included. The answers and/or the given dimensional units of all problems may be converted to metric units at the discretion of the instructor or student. Sooner or later you will be required to convert information to or from the metric system and to work completely with metric dimensions.

SUGGESTIONS FOR MATHEMATICAL OPERATIONS

One of Aesop's fables tells of a boy who cried for help when he ran into trouble swimming in deep water. A passerby loudly berated the boy for going in over his head. The boy, with undeniable logic, said, "Please help me now and scold me later."

As a technology student, you probably already have a better background in algebra and trigonometry than over half of the population. However, the possibility of error is always present. In the spirit of helpfulness, some of the more common errors are presented and corrected here; useful trigonometric formulas and a discussion of significant figures are also included.

An *equation* is a mathematical statement that two numbers are equal.

Examples: $4 + 8 = 3 \times 4$
$Y = X^2 - 4X + 3$

In the first equation above, the numbers on each side of the equal sign are obviously equal. The second equation is just as true but is not obvious.

A *formula* is an equation that expresses one physical quantity (or variable) in terms of other quantities, which themselves may vary.

Example: Formula for area of a circle: $A = \pi R^2$

There are a number of common mathematical errors that interfere with the solution and understanding of technical concepts. These errors often occur in the handling of fractions. Remember that when adding (or subtracting) fractions, all of the denominators must be the same. The numerators may then be added and the sum placed over the common denominator.

Example: $\dfrac{1}{16} + \dfrac{11}{16} - \dfrac{5}{16} = \dfrac{7}{16}$

The denominators must be made equal if they are not; but an important rule has to be remembered: *One times any number equals the number itself (including fractions).* For example, add the fractions

$$\dfrac{1}{2} + \dfrac{3}{4} - \dfrac{2}{3}$$

Our problem is to find a common denominator without changing the value of the individual fractions. The number 12 is chosen because 2, 3, and 4 are factors of 12. The denominator of each fraction is multiplied by the proper value to obtain 12. Since this operation by itself changes the value of the fraction, we must multiply the numerator by the same number. In this manner, each fraction has in effect been multiplied by 1.

$$\textit{Example:} \quad \frac{1}{2} \times \frac{6}{6} + \frac{3}{4} \times \frac{3}{3} - \frac{2}{3} \times \frac{4}{4} = \frac{7}{12}$$

The procedure to follow when dividing a number by a fraction is to *invert the denominator and multiply.*

$$\textit{Example:} \quad \frac{2}{3} \div \frac{4}{5} = \frac{2}{3} \times \frac{5}{4} = \frac{5}{6}$$

The reason is that the complete number would be simplified if the fraction could be removed from the denominator and replaced by the number 1. In order to do this, the denominator must be multiplied by its reciprocal, because any number multiplied by its reciprocal equals 1. Since the complete number cannot be changed, both the numerator and the denominator must be multiplied by the reciprocal of the denominator.

$$\textit{Example:} \quad \left(\frac{2}{3} \times \frac{5}{4}\right) \div \left(\frac{4}{5} \times \frac{5}{4}\right) = \frac{5}{6} \div 1 = \frac{5}{6}$$

Errors often occur when cancelling numbers in a fraction; confusion may arise because of failure to identify terms. A *term* is any part of an algebraic expression not separated by a plus (+) or a minus (−) sign. For example, a number may be designated by a single term composed of factors such as $32AB$. The factors in this case are three: 32, A, and B. Or, a number may be designated by two or more terms such as $12A + X - 3C$. The term $12A$ has two factors; X is one term; and $3C$ is a term with two factors. Remember that if there is only one term in the numerator and one term in the denominator, then like factors can be cancelled.

$$\textit{Example:} \quad \frac{48A^2B}{24A} = \frac{\cancel{24} \cdot 2 \cdot \cancel{A} \cdot A \cdot B}{\cancel{24A}} = 2AB$$

If there is more than one term in either the numerator or denominator, it is best not to try direct cancellation.

Example: $\dfrac{X^2 - 3}{X} \neq X - 3$

Instead: $\dfrac{X^2 - 3}{X} = \dfrac{X^2}{X} - \dfrac{3}{X} = X - \dfrac{3}{X}$

There are situations when the numerator or denominator may appear to have more than one term.

Example: $\dfrac{X(X^2 - 3)}{X} = X^2 - 3$

In this example, the numerator appears to have more than one term, but actually it has only one. X is one factor and $(X^2 - 3)$ is the other factor of the single term. Admittedly, this is confusing because the factor $(X^2 - 3)$ itself is composed of two terms. But these terms are enclosed in parentheses—a clue that you must be watchful.

Trigonometry

For the right triangle shown in Figure 1.1, the following formulas apply:

$$c = \sqrt{a^2 + b^2}$$

$$\text{sine } B = \dfrac{b}{c} = \dfrac{\text{opposite side}}{\text{hypotenuse}}$$

$$\text{cosine } B = \dfrac{a}{c} = \dfrac{\text{adjacent side}}{\text{hypotenuse}}$$

$$\text{tangent } B = \dfrac{b}{a} = \dfrac{\text{opposite side}}{\text{adjacent side}}$$

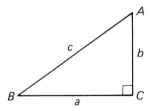

FIGURE 1.1 Right Triangle

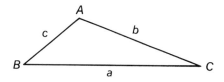

FIGURE 1.2 Triangle

For any given triangle, as shown in Figure 1.2, the following laws apply:

Sine law: $\dfrac{a}{\sin A} = \dfrac{b}{\sin B} = \dfrac{c}{\sin C}$

Cosine law: $a^2 = b^2 + c^2 - 2bc \cdot \cos A$

$b^2 = a^2 + c^2 - 2ac \cdot \cos B$

$c^2 = a^2 + b^2 - 2ab \cdot \cos C$

The sine, tangent, and radian values for small angles of 2° or less are equal to each other for at least four places to the right of the decimal. The values for 2° are all equal to 0.0349.

Significant Figures

It is a common practice to round off answers when the last digits are in doubt or not needed. Most of the answers in this text should be given to only three significant figures (or digits). The procedure to use when multiplying or dividing is to round the answer off to the number of significant figures in the least correct number used in the calculation, that is, the number with the least significant figures. For example, given the diameter of the earth as 7920 miles, determine the circumference. The figure given has three significant figures (the zero is not considered significant) and indicates that the diameter is known to within plus or minus five miles. Stated another way, we know the true diameter is somewhere between 7915 and 7925 miles. You must now decide on how many significant figures should be used for π. There would be no need to go beyond three significant figures (3.14). To demonstrate, we will use six significant figures for π.

$C = \pi D = 3.141\,59 \times 7920 = 24\,881.392\,80$ miles

This answer is misleading because it implies that we know the circumference to within plus or minus five one hundred thousandths of a mile. Actually, when the diameter is multiplied by π, the error

(± 5 miles) is also multiplied by π. Therefore, the answer can be correct only to three significant figures and should be rounded off to 24 900 miles. This means that the true circumference is known to be between the values of 24 850 and 24 950 miles.

When adding or subtracting, the place values of the significant figures are important.

Example: $384 + 4.2 + 0.006 + 49.68 = 437.886$

The number 384 is known to plus or minus 0.5; therefore, the answer should be rounded off to correspond. The answer should be written 438.

Sometimes there may be confusion as to the number of significant figures represented. Suppose the circumference of the earth is 24 900 miles, correct to five significant figures; how should this be indicated? Scientific notation is recommended; the figure can be written $2.4900(10)^4$ miles to indicate accuracy to five significant figures.

Procedure for Solving Problems

Step 1 Sketch a picture of the setup, and identify known and unknown quantities.

Step 2 Put the formula or formulas on the left-hand side of your paper, and substitute numerical values on the right. Perform the indicated operations, and write down the answer.

Step 3 If time is available, check your work several ways.

(a) Substitute rounded-off values to one significant figure, and see if your answer is comparable.

Example: $\dfrac{3.14(6.7)^2}{482} = 0.288$

Check: $\dfrac{3 \times 7^2}{500} = 0.3$

(b) Apply a dimensional check. Substitute dimensions in the formula, and see if the dimensions of the answer are correct.

Example: $\text{distance} = \dfrac{1}{2} \cdot g \cdot t^2$

$\text{feet} = \dfrac{\text{ft}}{\cancel{\text{sec}^2}}(\cancel{\text{sec}^2})$

(c) Ask yourself if the answer is reasonable, based on your experience. For instance, a calculated distance of 10 000 miles to the moon is obviously incorrect. Another example of an obvious error would be to obtain the diameter of a circle larger than the circumference.

Step 4 You should build your own list of formulas as you proceed through the text.

METRIC CONVERSION

The conversion to the metric or SI system of measurement is under way in the United States. You will no doubt be called upon by industry to make conversions to or from the U.S. Customary System (USCS) and the SI system, or be required to use data or charts in the SI system. SI stands for "The International System of Units." The official name in French is "Le Système International d'Unités." Most of the chapters in this book will have metric sample problems and metric student problems, in which you are expected to use all SI units in the formulas. A number of charts and tables in this text have the SI units contained in parentheses.

Most of us do not feel at home with the SI system. For example, if we are told that Boston is 20 miles away, we have a mental concept as to how far away that is in distance and time; but if we are told that Boston is 32 kilometres away, most of us have trouble visualizing whether that is next door or across the country. Practice will increase our familiarity with the SI system. A metric table is presented in the Appendix for your convenience.

There is one conversion that can be confusing. When comparing the masses of two or more objects, the habit of using weight (a force unit) has been established. In every supermarket you can find packages with the contents listed in ounces (force units) and grams (mass units). And just about every physics laboratory has a spring scale with ounce graduations on one side and grams on the other. Technically speaking, this should not be done, but when making comparisons it is convenient. The metric equivalent of the *pound* (lb) is the *newton* (N). The USCS equivalent of the *kilogram* (kg) is the *slug*. There are times when you must convert weight from mass units to force units or vice versa. For example, loads on bridges, beams, cranes, and hoists may be specified in kilograms (mass units); this information must be converted into newtons (force units) in order to use the required formulas when designing the structure.

2

Definitions, Equilibrium, Free Body Diagrams, and Concurrent Force Systems

OBJECTIVES

This chapter defines and explains common terms used in mechanics and strength of materials. Upon completion of this chapter, you will understand the meaning of the terms *free body diagram* and *equilibrium*. You will be able to construct a free body diagram (FBD), and you will be able to solve concurrent force problems.

INTRODUCTION

We have all laughed at animated cartoons where characters walk off a cliff but don't fall until they realize they are suspended in midair. But comparable situations are so common we rarely question them. For instance, why doesn't the middle of a bridge fall down? How are the forces directed to the supports? Why does Nature build tree limbs thicker at their bases than at their free ends? Exactly why is it hard to stand a dime on its edge? Why does the airplane's propeller turn and not the airplane? Why is it easier for your car to go up a slight grade rather than a steep one, even though the change in elevation is the same? The answers to these questions are based on mechanics.
  The term *mechanics* covers that area of physics dealing with forces acting on rigid bodies (you should review your physics text before beginning this chapter). A knowledge of mechanics, which is divided into statics and dynamics, is essential to the design of machines and structures. This text covers only *statics*.

DEFINITIONS

In order to understand force systems and their methods of solution, you must first learn the jargon. The terms used by physicists, engineers, and technicians to explain the force systems in this chapter and throughout the text are defined here. Also, a short definition of each type of force system is included for quick reference.

Equilibrium: An object (or body) is said to be in equilibrium when it is either at rest or moving in a straight line at constant velocity. Acceleration does not occur.

Statics: Statics is that part of mechanics dealing with force systems acting on rigid bodies that are in equilibrium. In most cases the bodies are at rest.

Dynamics: Dynamics covers that area of mechanics dealing with (1) forces producing a change in motion of an object and (2) the motions of objects without regard to the forces.

Rigid Body: A body is considered rigid if it does not deform when a force is applied. Actually, all materials deform to some extent under load, but for our applications we can consider structures and machine members as rigid bodies.

Scalar Quantity: A scalar quantity has magnitude only (not direction) and can be indicated by a point on a scale. Examples are temperature, mass, time, and dollars.

Vector Quantity: Vector quantities have magnitude and direction. Examples are wind velocity, distances between places on a map, and forces.

Force: A force is a vector quantity that, when applied to some rigid body, has a tendency to produce translation (movement in a straight line) or translation and rotation of the body. When problems are given, a force may also be referred to as a load or weight.

Reaction: The forces supplied by a structure's supports are referred to as reactions. The supports "react" to the weight of the structure and to applied loads, so that equilibrium is maintained.

Concentrated Force or Load: A concentrated force or load is a force that may be considered to be applied at one point.

Distributed Load: A distributed load is a force that is distributed along a line or over an area. The distances or areas are usually specified.

Definitions, Equilibrium, FBDs, and Concurrent Force Systems 11

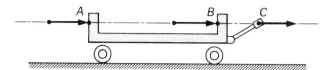

FIGURE 2.1 Line-of-Action

Line-of-Action: The line-of-action of a concentrated force is the infinite extension of the line along which the force acts. A force, or its components, may be moved along the force's line-of-action for convenience in problem solutions. As an example, the wagon in Figure 2.1 will react the same way to the applied horizontal force, whether or not the point of application is at A, B, or C.

Collinear: If several forces lie along the same line-of-action, they are said to be collinear. It makes no difference whether the forces oppose each other or not.

Coplanar: When all forces acting on a body are in the same plane, the forces are coplanar. If the forces do not lie in the same plane, they are noncoplanar. All of the force systems discussed in this text will be coplanar or in parallel planes.

Concurrent Force Systems: A concurrent force system contains forces whose lines-of-action meet at some one point. Concurrent force systems are illustrated in Figure 2.2. Forces may be *tensile* (pulling), as in Figure 2.2(a), or may be *compressive* (pushing), as in Figure 2.2(b).

Moment of a Force: The measure of the effect of a force tending to produce rotational motion or a twisting action is the moment of that force. As illustrated in Figure 2.3, a moment of a force is taken about an axis Y-Y, which is perpendicular to the plane in which the force lies. A moment (M) is the product of the magnitude of a force (F) and the perpendicular distance (D) to the moment center:

$$M = F \times D$$

The point in the plane through which axis Y-Y passes is called the *moment center* and is labeled O in the figure. The units of measurement used in this text will usually be in pound·inches (lb·in.), pound·feet (lb·ft), or newton·metres (N·m).

Torque: The moment of a force may be referred to as torque. The term *torque* is preferred to *moment of force* when discussing rotating shafts and equipment or when discussing the effects of tight-

CHAPTER TWO

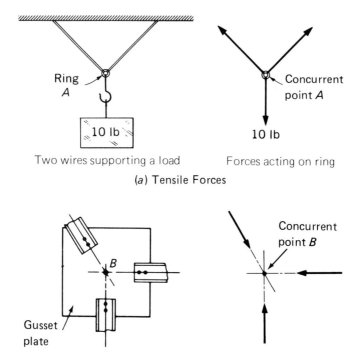

FIGURE 2.2 Concurrent Force Systems

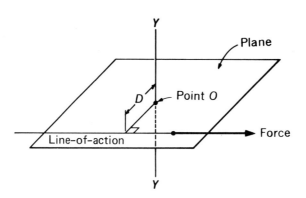

FIGURE 2.3 Moment of a Force about Axis Y-Y

Definitions, Equilibrium, FBDs, and Concurrent Force Systems   13

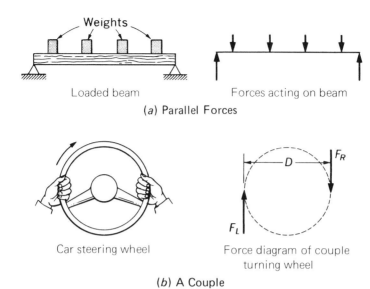

FIGURE 2.4   Parallel Force Systems

ening nuts and bolts. The effects of torque when applied to shafts are discussed in Chapter 12 on torsion. Torque is measured in the *moment* terms of lb·in., lb·ft, or N·m.

Parallel Force System: Just as the name implies, all forces are parallel. See Figure 2.4(*a*) and (*b*).

Couples: A single force acting on a body cannot produce rotation of the body without also producing translation. Therefore, a force system known as a *couple* may be needed to produce rotation *only*. A couple is a special case of a parallel force system and consists of one pair of forces equal in magnitude but opposite in direction, with the lines-of-action parallel. The *moment* is the measure of the magnitude of a couple. For instance, "The couple produced a moment of 800 lb·ft." Refer to Figure 2.4(*b*).

Nonconcurrent Force System: A nonconcurrent force system consists of forces whose lines-of-action do not intersect at one point. Parallel force systems and couples are nonconcurrent, but this text will apply the term *nonconcurrent* to those systems that are neither parallel nor couples. Figure 2.5 shows a nonconcurrent force system.

Sign Convention: All forces heading to the right (horizon-

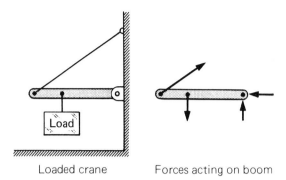

Loaded crane   Forces acting on boom

FIGURE 2.5 A Nonconcurrent Force System

tally) or up (vertically) will be given a plus (+) sign. Forces heading down (vertically) or left (horizontally) will be given a minus (−) sign.

All moments (see definition above) with a clockwise (CW) rotation about the moment center (↻) will be given a plus (+) sign. All counterclockwise (CCW) moments (↺) will be given a minus (−) sign.

One technique some technicians and mathematicians like to use concerns the situation where you may have to *assume* the direction of a force in order to solve the problem. If your answer comes out negative (−), it means you assumed the wrong direction. This is a normal method of procedure and does not mean you did the problem wrong.

FREE BODY DIAGRAMS

Before proceeding further in mechanics, you should become familiar with a device used to aid in the solution of problems. This device is called a *free body diagram*. You may have noticed that Figures 2.2, 2.4, and 2.5 included sketches showing forces as vectors. These sketches are called free body diagrams (FBDs). An FBD is a sketch of a structure or part of a structure with all connecting parts removed. The removed parts are replaced with vectors indicating the applied forces and their directions. For example, in Figure 2.6 we are given a sketch of a ladder leaning against a wall and supporting a painter. Our problem is to determine the reaction forces applied to the ladder. It would help if everything touching the ladder could be stripped away in order to get a clear picture of the forces being applied—that is, free the ladder from all contacting objects. The ladder is sketched in the proper position but the wall, ground, and painter are not shown. Instead, the painter

Definitions, Equilibrium, FBDs, and Concurrent Force Systems   15

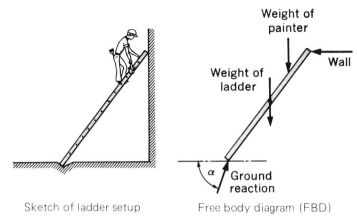

FIGURE 2.6   Example of a Free Body Diagram

is replaced with an arrow acting downward to indicate the weight of the painter. The wall is replaced by a horizontal arrow indicating the force exerted by the wall against the ladder. The weight of the ladder itself is shown by an arrow acting downward through the center of gravity of the ladder. The ground is replaced by an arrow at some still-unknown angle to the horizontal to indicate the force that not only prevents the ladder from slipping but also supports the combined weight of the ladder and painter. This picture is called a free body diagram. Since the FBD is drawn prior to the solution of a problem, a number of force values and directions will not be known. In such a situation, the force directions are assumed and corrected later if wrong.

Frequently, supports for structures are deliberately designed to apply either (1) a resisting force in only one direction, or (2) a resisting force in any direction but not a resisting moment, or (3) a resisting force and a resisting moment. Therefore, in order to indicate the direction in which forces or reactions can be applied, various standard symbols are employed in the original sketch of a structure. The most common symbols are shown in the left-hand column of Figure 2.7. Corresponding FBDs are shown in the right-hand column.

1. A roller can support a force only in a direction normal to the surface it is resting on. Any force in another direction will cause it to roll. In the actual structure, a rocker may be used in place of the roller. Also, knife edges are frequently used in illustrations to indicate a concentrated force in the direction of the knife edge. See Figure 2.7(a) and (b).
2. A plane surface is considered frictionless unless otherwise spec-

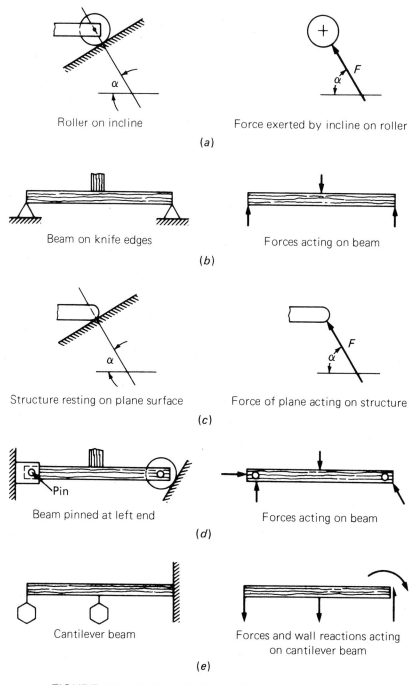

FIGURE 2.7 Various Methods of Supporting a Structure

ified. Therefore, a plane surface can support a load only in a direction normal to its surface. See Figure 2.7(c).
3. Pins or bolts used to connect two members of a structure or machine can resist forces in all directions perpendicular to their longitudinal axes, but they cannot resist couples. As a demonstration, drill a small hole in one end of a metre stick. Place a nail through the hole and support the metre stick by holding onto the nail. It is obvious that the nail (our so-called pin) will support the metre stick in a vertical position. The nail will not support the metre stick in a horizontal position because of the moment applied by the metre stick's weight acting through the center of gravity. See Figure 2.7(d). Note that the components of the force at the pin location are shown.
4. If a beam is firmly embedded in some surface (a cantilever beam), the supporting structure can resist with forces and couples. See Figure 2.7(e).

As we go through the sample problems in the text, notice the original sketches of the setups and the related FBDs. The FBD concept is a versatile tool and can be used to include complete machine assemblies or structures, or any number of parts of a structure. Also, a structure may be separated into its individual members, and the members themselves may be cut into sections to evaluate the forces present. Keep in mind two points:

1. The free body diagram is drawn only to help in the solution of a problem and therefore must be the first step in a solution. As the solution develops, new information can be added to the diagram.
2. The FBD is used to indicate the forces acting *on* the body and does *not* show the forces the body applies to other objects in contact.

MECHANICAL EQUILIBRIUM

Let us now continue to another topic in mechanics, *mechanical equilibrium*. We are all mystified by the magician who levitates a woman, because we realize that some force is needed to oppose gravity. Also, everyone realizes that when we sit in a chair, the chair must push back. But do we all understand that the chair pushes against us with exactly the same force as our weight? Do we recognize that if a larger person sits in the chair, it will push against that person with exactly his or her weight? To investigate this and more complicated situations, we must start with Newton's First Law of Motion:

> *If an object is at rest, it will remain at rest, or if it is in motion, it will remain in motion at a constant speed and in a straight line, unless some unbalanced force changes its condition.*

When either of the above conditions exists, the object is said to be in equilibrium. Newton was saying, in effect, that when an object is in equilibrium (1) there are neither forces nor couples applied, or (2) the vectorial sum of all forces and the sum of all moments equal zero. An alternative way to state these conditions is to say that the following three static equilibrium equations must be satisfied:

1. The sum of all the forces in the X direction must equal zero.

    $\Sigma F_x = 0$

2. The sum of all the forces in the Y direction must equal zero.

    $\Sigma F_y = 0$

3. The sum of all the moments (about any moment center on or off the object) must equal zero.

    $\Sigma M = 0$

These equations, representing the whole coplanar concept of statics, will be useful in our study of various types of force systems used to maintain a structure in equilibrium.

CONCURRENT FORCE SYSTEMS

The moment equation ($\Sigma M = 0$) is not needed for concurrent force problems, because all of the forces have a zero moment about the point of concurrency. The equations needed are those required to sum the forces in the X and Y directions: $\Sigma F_x = 0$, and $\Sigma F_y = 0$. Two methods of solution may be used:

1. A force polygon is sketched and the unknown values calculated; or, the polygon can be drawn to scale and solved graphically.
2. The X and Y components of the forces can be summed, in accordance with Newton's First Law.

Definitions, Equilibrium, FBDs, and Concurrent Force Systems 19

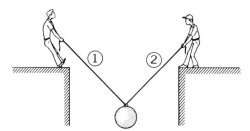

FIGURE 2.8 Concurrent Force Demonstration

Let us analyze the situation shown in Figure 2.8. Here, two people are slowly lifting a 120 lb weight, and we want to know how much force each person must exert. The force polygon method of solution is based on the fact that if a number of forces maintain an object in equilibrium, the force polygon will close. That is, when you are constructing the force polygon one force at a time, the head of the last force constructed will end at the starting point.

Procedure for Force Polygon Method

Step 1 Draw an FBD of the point of concurrence.

Step 2 Either draw to scale or sketch the force polygon. Start with the known force or forces. Put the tail of one vector at the head of the preceding vector. Draw the line-of-action of the unknown forces at the proper angles and in such a manner that the polygon closes (the last unknown force must return to the starting point).

Step 3 Use trigonometry to solve for the forces, or if drawing to scale, measure the lengths of the forces and convert to force units.

Sample Problem 2.1 uses the force polygon method to solve the problem illustrated in Figure 2.8.

SAMPLE PROBLEM 2.1
Concurrent Forces (Force Polygon Method)

Problem

Figure 2.8 shows two people lifting a 120 lb weight. Each rope makes an angle of 45° with the horizontal. What is the force in each rope?

Solution

Step 1 Draw the FBD of the point of concurrence—that is, the ring at the top of the weight.

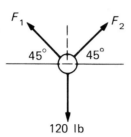

FBD of ring on weight

Step 2 (a) Using the FBD as a guide, start your force polygon (in this case, a triangle) by drawing the known force vector.

(b) Construct F_1 next (F_2 could have been chosen). The magnitude is not known, but its direction and angle to the horizontal are known. The line-of-action of F_1 is drawn from the head of the known vector.

(c) The lengths of vectors F_1 and F_2 are not known, but we know that F_2 has to get us back to the starting point. The F_2 line-of-action is drawn at a 45° angle through the starting point. Arrowheads are now placed on the forces and the angles of the triangle determined.

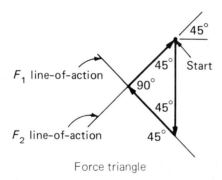

Force triangle

Step 3 Since the force triangle is determined to be a 45°–90° triangle:

$$F_1 = F_2 = 120 \text{ lb} \times \cos 45° = 85 \text{ lb}$$

The problem represented by Figure 2.8 will be solved again by summing the X and Y forces. However, there are extra steps involved, because two forces are at angles to the X and Y axes. Therefore, we

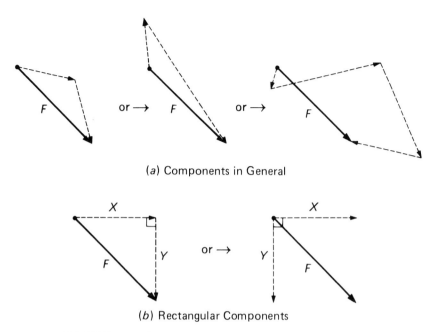

FIGURE 2.9 Components of a Force

must solve for the components of these forces first and then solve for the forces. The word *component* refers to two or more forces whose vectorial sum is equal to a given force. Notice that in Figure 2.9(a) a force can have any number of components in any number of directions. For our purposes, though, it is more convenient to break a force down into just two rectangular components. Note that the word *rectangular* means the components are perpendicular to each other and *not* perpendicular to the force they represent. Components are most commonly placed in the X and Y directions, as illustrated in Figure 2.9(b).

As Newton indicates, the X and Y components for any number of forces can be added together algebraically (X to X and Y to Y). So, we use Newton's equations when summing components to determine the missing values. When this is complete, we then use the two components of each unknown force to determine the magnitude (and possibly direction) of that force.

Before discussing the procedure for using Newton's equations, we must discuss the method of determining the magnitude and direction of a force from its components (vector summation). This method makes use of the Parallelogram Law:

> *A force is the diagonal vector of the parallelogram for which the components are the adjacent sides.*

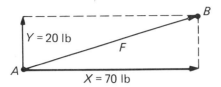

FIGURE 2.10 Illustration of the Parallelogram Law

Refer to Figure 2.10. Let the Y component of an unknown force equal 20 lb and the X component equal 70 lb. Place the tails of the components together, as at point A in the figure. Draw parallel lines to the components from the heads of each component. The unknown force has the magnitude and direction represented by the diagonal from A to B in the figure. Thus:

$$F = \sqrt{20^2 + 70^2} = 72.8 \text{ lb}$$

In your efforts to understand the concept of components of a force, it may help to remember that the components could replace the actual force with no change in effect.

Now we are ready to use Newton's equations: $\Sigma F_x = 0$ and $\Sigma F_y = 0$. Three important points must be kept in mind when summing X and Y components. These points, which refer specifically to the situation where a vertically applied load is supported by two unknown forces, are as follows:

1. The vertical components (Y_1 and Y_2) support the load.
2. The horizontal components (X_1 and X_2) do not support the load but oppose each other and therefore must be equal and opposite in direction. If this were not so, the load would move sideways.
3. If the angles the forces make with the horizontal are equal, then $Y_1 = Y_2$ and $F_1 = F_2$. That is, the forces share the load equally. If the angles are not equal, then you must use the fact that $X_1 = X_2$ to help you solve the problem.

Procedure for Component Summation Method

Step 1 Draw the FBD of the point of concurrence.

Step 2 Sum the X and Y components, and set them equal to zero.

Step 3 Apply simultaneous equation methods and the Parallelogram Law to solve for the unknown force or forces.

Sample Problems 2.2 and 2.3 illustrate the component summation method for solving concurrent force problems.

SAMPLE PROBLEM 2.2
Concurrent Forces (Component Summation Method)

Problem

Figure 2.8 shows two people lifting a 120 lb weight. Each rope makes an angle of 45° with the horizontal. What is the force in each rope?

Solution

Step 1 Draw the FBD of the concurrent point.

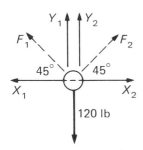

FBD of ring on weight

Step 2 Sum the X and Y components:

(a) $\Sigma F_x = 0 = -X_1 + X_2$
(b) $\Sigma F_y = 0 = Y_1 + Y_2 - 120$ lb

Step 3 Rewrite the preceding equations and solve:

(a) $0 = -F_1(\cos 45°) + F_2(\cos 45°) = -0.707 F_1 + 0.707 F_2$
(b) $0 = F_1(\sin 45°) + F_2(\sin 45°) - 120$ lb
$ = 0.707 F_1 + 0.707 F_2 - 120$

Add (a) and (b):

$1.414 F_2 - 120 = 0$
$F_2 = 85$ lb

Substitute in (a):

$F_1 = \dfrac{0.707 \times 85}{0.707} = 85$ lb

SAMPLE PROBLEM 2.3
Summation of X and Y Components (Concurrent Forces in Equilibrium)

Problem

Given the information in the following sketch, determine the force in each wire.

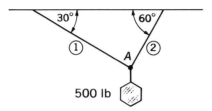

Solution

Step 1 Draw the FBD of point A.

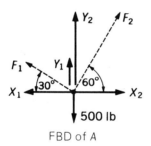

FBD of A

Step 2 Sum the X and Y components:

(a) $\Sigma F_y = 0 = -500 + Y_1 + Y_2$
$ = -500 + F_1(\sin 30°) + F_2(\sin 60°)$

(b) $\Sigma F_x = 0 = -X_1 + X_2 = -F_1(\cos 30°) + F_2(\cos 60°)$

Step 3 Rewrite the preceding equations and subtract (b) from (a):

$0.5F_1 + 0.87F_2 - 500 = 0$
$\underline{-(-1.5F_1 + 0.87F_2 = 0)} \quad \text{(multiplied by 1.74)}$
$2.0F_1 - 500 = 0$

$F_1 = 250 \text{ lb}$

Substitute in (b):

$$F_2 = \frac{250 \, (\cos 30°)}{\cos 60°} = 432 \text{ lb}$$

Before starting the end-of-chapter problems, there is one more statement that may be helpful in solving some problems:

If just three nonparallel forces are maintaining a body in equilibrium, they must be concurrent.

You may want to check this statement by trying various combinations of forces that are not concurrent. Resolve any two forces into a single force (consider the two forces components of the single force). This single force must be collinear and equal and opposite to the third force for equilibrium. Sample Problem 2.4 applies the above principle.

SAMPLE PROBLEM 2.4
Three Concurrent Forces in Equilibrium

Problem

In the following sketch, the length of the ladder is 40 ft and the man is standing 30 ft up the ladder. Consider the ladder weightless. Solve for the ground and wall forces acting on the ladder.

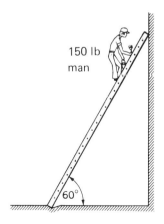

26 CHAPTER TWO

Solution

Step 1 Draw the FBD of the ladder. Sketch in the point of concurrence O and determine the geometry of the forces.

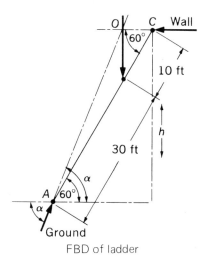

FBD of ladder

vertical height of C above A = 40 ft × sin 60° = 34.7 ft
horizontal distance A to C = 40 ft × cos 60° = 20 ft

We know that point O must be on the 150 lb vector and that the wall reaction is horizontal. Therefore:

distance O to C = 10 ft × cos 60° = 5.0 ft

$$\alpha = \tan^{-1}\left(\frac{34.7}{20 - 5.0}\right) = 66.6°$$

Step 2 We can now draw the force triangle.

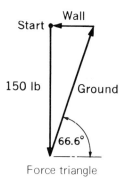

Force triangle

Step 3 Solve for the reaction forces.

$$\text{ground force} = \frac{150}{\sin 66.6°} = 163 \text{ lb}$$

$$\text{wall force} = \frac{150}{\tan 66.6°} = 65 \text{ lb}$$

SUMMARY

A free body diagram is used to indicate the forces acting *on* the body and does *not* show the forces the body is exerting on the other machine members or structures in contact with it.

If a body is in equilibrium, the following three static equilibrium equations must be satisfied:

$$\Sigma F_x = 0$$
$$\Sigma F_y = 0$$
$$\Sigma M = 0$$

Because the summation of X and Y force components are zero when a body is in equilibrium, the force polygon representing the forces will close.

Concurrent force systems contain forces whose lines-of-action meet at some single point (the point of concurrence). Solutions to concurrent force problems can be obtained by the force polygon method or the summation of components method.

PROBLEMS

Concurrent Force Systems

Draw FBDs for each problem. Unless otherwise specified, angles are measured in a counterclockwise direction from the $+X$ axis.

1. Figure 2.11 shows a group of people in a tug-of-war. The arrows indicate the direction each person is pulling on the rope. If each person can pull with 150 lb of force, what is the resisting force in the rope at sections *AA*, *BB*, *CC*, and *DD*?

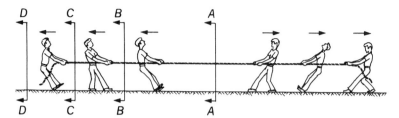

FIGURE 2.11 Tug-of-War for Problem 1

2. A girl places two bathroom scales together, one on top of the other, and then stands on them. She weighs 100 lb. What is the reading on each scale? Ignore the weight of the scales.
3. We are given two components. The 50 lb component acts vertically down, and the 248 lb component acts horizontally to the left. Solve for the resultant force.
4. Two components pull away from the point of concurrence. The 100 lb component acts at an angle of 100°, and the 150 lb component acts at an angle of 140°. Solve for the resultant force.
5. A 320 lb force is directed upward at an angle of 50°. Solve for the vertical and horizontal components.
6. A 1000 lb force is directed upward at an angle of 160°. Solve for the X and Y components.
7. A force is directed upward at an angle of 35°. Its Y component is +300 lb. Determine the magnitude of the force and the magnitude and direction of the X component.
8. A 4000 lb horizontal force is directed toward a surface that is inclined at an angle of 50°. Determine the components of the force that are parallel and normal (perpendicular) to the inclined surface.
9. Refer to Figure 2.12. Determine which set of force systems is in equilibrium.
10. For those force systems in Figure 2.12 that are *not* in equilibrium, determine the balancing force required to place the body in equilibrium.
11. Refer back to Figure 2.8. We are given that the weight equals 50 lb, the angle of rope 1 is 135°, and the angle of rope 2 is 60°. What is the force in each rope?
12. Given the information in Figure 2.13, determine the force in each supporting wire.
13. The stoplight in Figure 2.14 is supported by two wires. The light weighs 75 lb, and the wires make an angle of 10° with the horizontal. What is the force in each wire?

Definitions, Equilibrium, FBDs, and Concurrent Force Systems   29

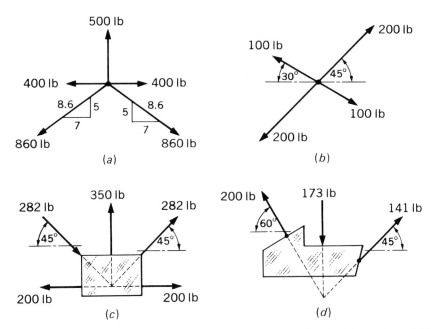

FIGURE 2.12   Concurrent Force Systems for Problems 9 and 10

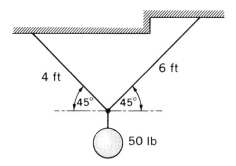

FIGURE 2.13   Problem 12

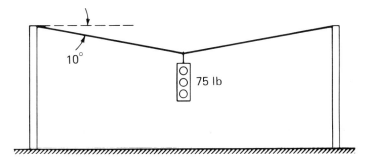

FIGURE 2.14   Problem 13

14. Refer to Figure 2.15. What horizontal force F is required for equilibrium?

15. Refer to Figure 2.16. This illustrates a boom and wire supporting a load. The boom exerts a horizontal force at the point of concurrence of the boom, wire, and load. Determine the forces in the boom and the wire.

16. The roller in Figure 2.17 weighs 1200 lb. Determine the force F required for equilibrium in sketches (a) and (b).

17. Refer to Figure 2.18. Determine the force in each of the four wires.

18. Figure 2.19(a) shows a 50 lb block resting on a frictionless incline. Solve for the forces F_1 and F_2 required for equilibrium.

19. Figure 2.19(b) shows a 70 lb block resting on an incline. Friction between the block and incline requires F_2 to be positioned as shown. Solve for F_1 and F_2 required for equilibrium.

20. Refer to Figure 2.20. Length A = 5 ft, length B = 10 ft, and angle α = 30°. Determine angle β of the incline in order to maintain equilibrium.

21. Refer to Figure 2.21. Solve for the resisting force at pin A to maintain equilibrium.

22. (SI Units) A 30 N force is directed down at an angle of 225°. A 40 N force is directed down at an angle of 315°. Consider these forces components, and solve for the resultant force.

23. (SI Units) Solve Problem 11 if the load is 100 kg. Give the answers in newtons (N).

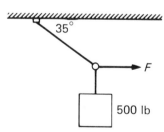

FIGURE 2.15 Problem 14

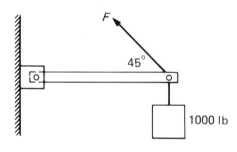

FIGURE 2.16 Problem 15

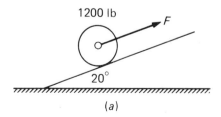

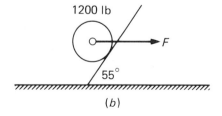

(a)   (b)

FIGURE 2.17 Problem 16

Definitions, Equilibrium, FBDs, and Concurrent Force Systems   31

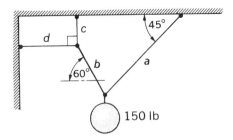

FIGURE 2.18   Problem 17

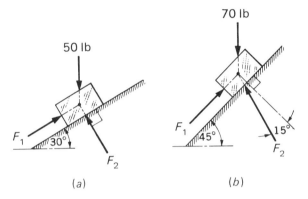

FIGURE 2.19   Problem 18

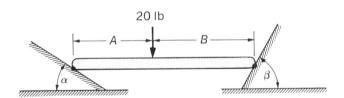

FIGURE 2.20   Problem 20

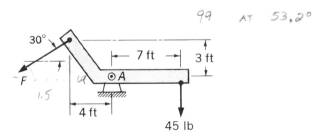

FIGURE 2.21   Problem 21

COMPUTER PROGRAM

The following program will solve for two forces required to balance a third known force; the setup is similar to that shown in Figure 2.8 or explained in Sample Problem 2.3. The forces must all be either pulling away from the point of concurrence or all pushing toward it.

```
10   REM  FORCE(2)
20   PRINT "THIS PROGRAM WILL SOLVE FOR TWO UNKNOWN FORCES IN A CONCURRENT
     FORCE SYSTEM WITH ONE KNOWN APPLIED FORCE."
25   PRINT
30   PRINT
40   PRINT "ALL ANGLES ARE MEASURED AT THE POINT OF CONCURRENCY AND IN A C
     CW DIRECTION FROM THE - PLUS X - AXIS."
50   PRINT
60   PRINT " ENTER KNOWN FORCE."
70   INPUT F
90   PRINT "ENTER ANGLE OF KNOWN FORCE."
100  INPUT A
105  PRINT
120  PRINT "ENTER ANGLE OF FORCE #1."
130  INPUT B
150  PRINT "ENTER ANGLE OF FORCE #2."
160  INPUT C
180  LET D = ( COS (3.14 / 180 * B) * SIN (3.14 / 180 * C)) - ( SIN (3.1
     4 / 180 * B) * COS (3.14 / 180 * C))
190  LET F1 =  INT ((F * SIN (3.14 / 180 * A) * COS (3.14 / 180 * C) -
     F * COS (3.14 / 180 * A) * SIN (3.14 / 180 * C)) / D * 10 + .51) /
     10
200  LET F2 =  INT (( - F * COS (3.14 / 180 * B) * SIN (3.14 / 180 * A)
      + F * COS (3.14 / 180 * A) * SIN (3.14 / 180 * B)) / D * 10 + .51
     ) / 10
210  PRINT
220  HOME
230  PRINT    TAB( 15)"KNOWN"; TAB( 25)"F1"; TAB( 35)"F2"
240  PRINT    TAB( 15)"FORCE"
250  PRINT
260  PRINT "ANGLES =";  TAB( 15);A; TAB( 25);B; TAB( 35);C
270  PRINT
280  PRINT "FORCES = ";  TAB( 15);F; TAB( 25);F1; TAB( 35);F2
290  END
```

3

Nonconcurrent Force Systems

OBJECTIVES

This chapter groups the various force systems (other than concurrent) into categories for easier identification and solution. Upon completion of this chapter, you will be able to (1) identify the type of force system acting on a structure and (2) replace this system with an equivalent system or supply balancing forces to obtain equilibrium.

INTRODUCTION

Later in this text, you will be deciding how large a beam, column, or machine part is needed to support a load. Before a decision can be made, all of the forces acting on the structure must be determined. This chapter, in addition to Chapter 2, provides information needed to help design and evaluate machines and structures.

MOMENT OF A FORCE

A knowledge of moments is required to solve nonconcurrent force systems. Remember, Newton's three equations for equilibrium included $\Sigma M = 0$. In Chapter 2, a moment was defined as the measure of a twisting action. The value of a moment is the product of a force and the perpendicular distance from the line-of-action of the force to the moment center. The formula is as follows:

$$M = F \times D$$

34   CHAPTER THREE

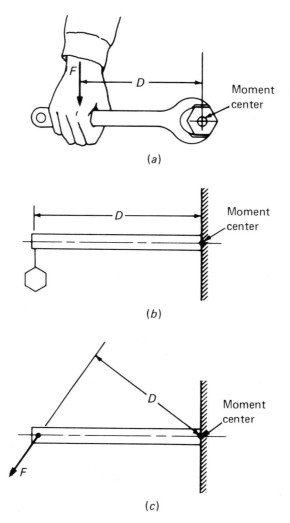

FIGURE 3.1   Moment Applications

The distance D is frequently called the *moment arm* or *torque arm*. The moment center may or may not be specified in a problem. If it is not specified, its position may be apparent (as the axle of a pulley), or the position may be any convenient location you choose.

A familiar application of moments is the process of removing the lug nuts from the wheel of a car. See Figure 3.1(a). If the nut is on too tight, what do most of us do? We try to get a wrench with a longer handle. Since we have reached the limit of the force that we can apply, we are trying to increase the moment (or torque) by increasing D. The torque the wrench applies to the nut is opposed by the forces preventing the nut from twisting. So, we must apply a small extra amount of torque above the maximum the nut can resist.

Cantilever beams (beams supported at only one end) support loads, which apply moments at the wall. See Figure 3.1(b). And, of course, the wall must supply a balancing moment to provide equilibrium. Now, what about the situation illustrated in Figure 3.1(c)? The line-of-action is not parallel to the wall, so we must calculate the perpendicular distance to the moment center. However, there is another method devised by Pierre Varignon (1654–1722), a French mathematician. This method may be easier to use in some cases. Varignon proved a useful fact now known as Varignon's Theorem:

> *The moment of a force is equal to the sum of the moments of its components.*

Varignon's Theorem is demonstrated in Sample Problem 3.1. The transfer of torque between mating gears in the sample problem represents an application similar to that of the cantilever beam shown in Figure 3.1(c).

SAMPLE PROBLEM 3.1
Moment of a Force

Problem

The following illustration shows the force the driver gear applies to the driven gear. The 40 lb force is applied at an angle of 20° to the horizontal. What is the moment of the force (torque) applied to the driven gear?

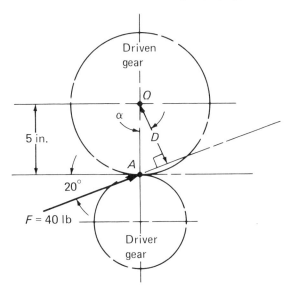

Solution with Distance D

The moment center is at O and angle α is 20° because its sides are mutually perpendicular to the sides of the 20° angle shown. Therefore:

$D = 5 \times \cos 20° = 4.70$ in.
$M = F \times D = 40$ lb $\times 4.70 = 188$ lb·in. CCW

Solution with Varignon's Theorem

The moment center is at O. The components will be placed at point A because (1) A is on the line-of-action of the force and (2) the Y component need not be calculated, because it has no moment arm and its moment effect is zero. Therefore:

$X = 40$ lb $\times \cos 20° = 37.6$ lb
$M = F \times D = 37.6$ lb $\times 5$ in. $= 188$ lb·in. CCW

PARALLEL FORCE SYSTEMS

We will consider parallel force systems acting on a cantilever beam first. See Figure 3.2(a). All of the reactions are at the wall because that is the only support. Notice in Figure 3.2(b) the FBD of the beam shows a vertical force at the wall; we will refer to it as the *direct force*. This force must balance the applied forces:

$\Sigma F_y = 0 = -$ applied force + direct reaction force at wall

The FBD also shows a curved arrow at the wall. This represents a reaction couple that balances the applied moment:

$\Sigma M = 0 = -$ applied moment + reaction couple at wall

You may be asking why a couple is specified at the wall and not a moment. This is done to separate the force systems, so that the couple forces won't be confused with the other wall reaction force. Also, the couple forces can be applied in a number of different ways, and at this stage of design we are interested in the reactions, not how they will be applied. For example, look at Figure 3.2 again. This figure shows a cantilever beam 5 ft long and supporting a 100 lb load. The reactions at the wall are:

$\Sigma F = 0 = -100$ lb $+ F_w$
$F_w = +100$ lb

Nonconcurrent Force Systems 37

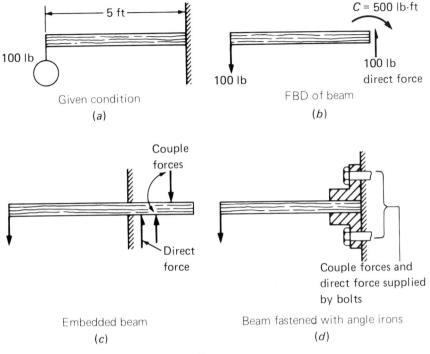

FIGURE 3.2 Cantilever Beam

$$\Sigma M = 0 = -(100 \times 5) + C_w$$
$$C_w = +500 \text{ lb·ft}$$

This information tells us that the wall and the beam must be strong enough to support a force of 100 lb and resist a moment equal to 500 lb·ft. Two ways the wall can do this are shown in Figure 3.2. The beam can be lengthened and embedded in the wall as shown in Figure 3.2(c). The resisting forces making up the couple are distributed along the embedded beam, but notice that now there is a force acting *down* on the beam in the wall. Figure 3.2(d) shows that the beam might be fastened to the wall with angle irons and bolts placed horizontally. The forces the bolts must resist to supply the couple will vary depending on how far apart they are.

So, you can see that we don't want to get involved yet in how the beam will be supported. For the present we only want to know the reactions.

The moment center for cantilever beams is taken at the wall. For beams on two supports, the procedure used to solve for the reactions is a bit different.

Procedure for Finding Reactions of Beams on Two Supports

Step 1 Draw an FBD.

Step 2 Sum the moments of all the forces on the beam about a moment center taken at one of the supports. This will provide you with the reactive force at the other support.

Step 3 Sum moments about the second support to determine the reactive force at the first support.

Step 4 As a check, sum the applied and reaction forces. These should equal zero.

Sample Problem 3.2 illustrates the solution of parallel forces applied to a beam on two supports.

SAMPLE PROBLEM 3.2
Parallel Forces
Problem

The beam in the following sketch is supported by a pin at A and a roller at B. Given the distributed loads in the sketch, determine the resisting forces at A and B.

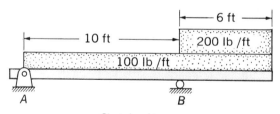

Sketch of beam

Solution

Step 1 Simplify by replacing the distributed loads with single forces at their centers of gravity, as indicated in the FBD.

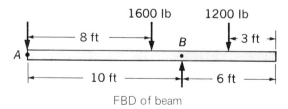

FBD of beam

Step 2 Sum moments about a moment center placed at A.

$$\Sigma M_A = 0 = +1600 \text{ (8 ft)} + 1200 \text{ (13 ft)} \pm F_B \text{ (10 ft)}$$

Note: The applied forces result in a + clockwise moment; therefore the resisting moment must be − counterclockwise.

$\Sigma M_A = 0 = +28\,400 - F_B (10)$

$F_B = 2840$ lb ↑

Step 3 Sum moments about a moment center placed at B.

$\Sigma M_B = 0 = +1200\,(3\text{ ft}) - 1600\,(2\text{ ft}) \pm F_A\,(10\text{ ft})$

$F_A = 40$ lb ↓

Note: The resisting force F_A is *down* in order to provide a counterclockwise moment about B.

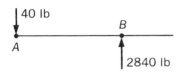

Step 4 Sum the applied and reaction forces.

Check: $\Sigma F_y = 0 = -40 - 1600 + 2840 - 1200$

METRIC SAMPLE PROBLEM 3.3
Parallel Forces

Problem

Given the information in the following sketch, solve for the reactions at A and B. Remember, mass units (kilograms) must be converted to force units (newtons).

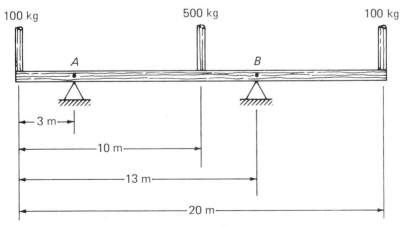

Sketch of beam with loads

40 CHAPTER THREE

Solution

Step 1 Convert mass units to force units and draw the FBD.

$$100 \text{ kg} \times 9.81 = 981 \text{ N}$$
$$500 \text{ kg} \times 9.81 = 4905 \text{ N}$$

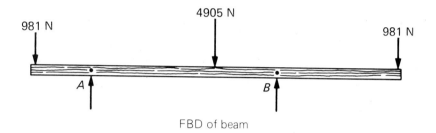

FBD of beam

Step 2 Sum the moments about one support (support A is chosen).

$$\Sigma M_A = 0 = -(981 \times 3) + (4905 \times 7) + (981 \times 17) - (B \times 10)$$
$$0 = -2943 + 34\,335 + 16\,677 - 10B$$
$$B = \frac{48\,089}{10} = 4809 \text{ N} \uparrow$$

Step 3 Sum the moments about B.

$$\Sigma M_B = 0 = -(981 \times 13) - (4905 \times 3) + (981 \times 7) + (A \times 10)$$
$$0 = -12\,753 - 14\,715 + 6867 + 10A$$
$$A = \frac{20\,601}{10} = 2060 \text{ N} \uparrow$$

Step 4 Check to three significant figures.

Check: $\Sigma F = 0 = -981 - 4905 - 981 + 4809 + 2060$

PULLEYS

Pulleys may be analyzed as either parallel force systems or concurrent force systems, depending on the application of the loads, as illustrated in Figure 3.3. A pulley is useful because it cannot resist an applied moment. To demonstrate this statement, let us refer to Figure 3.3(a). The pulley is free to rotate about its hub A. A rope running over the pulley has a 100 lb load placed on it. From your past experience, you would probably assume (correctly) that the load must be balanced by

Nonconcurrent Force Systems   41

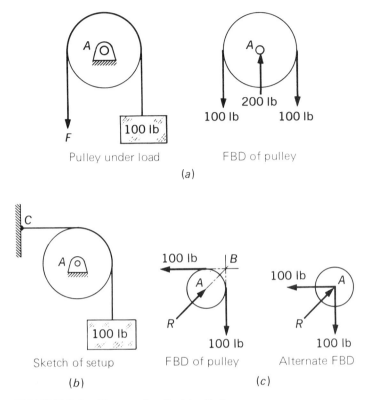

FIGURE 3.3 Forces Applied to Pulleys

a 100 lb force on the other end of the rope. The reason is that a summation of moments about the hub at A must equal zero for the pulley to be in equilibrium. If the summation were not zero, the pulley would start to rotate and continue to increase its angular velocity. Also, the forces on each end of the rope must be the same, because their moment arms are the same (the pulley radius).

Note that in Figure 3.3(a) we have a parallel force system. The fact that the forces on each side of the pulley must be equal aids us in solving multiple pulley problems. The procedure, as demonstrated by Sample Problem 3.4, is to draw an FBD of one of the multiple pulleys, usually the movable one. The next step is to determine the force on each rope section by summing forces. The force on any one section must equal the force on any other; thus we can solve our problem.

A somewhat different pulley situation is shown in Figure 3.3(b). Knowing that the sum of the moments about A (the hub) has to be zero, it is easy to determine that the force exerted by the hook at C is 100 lb. The forces are concurrent at point B, as indicated in the FBD. However, it is frequently more convenient to show the forces applied to the hub, as in Figure 3.3(c). This can be done because the sum of the moments about A is zero, just as it is at B.

SAMPLE PROBLEM 3.4
Pulleys in Equilibrium

Problem

Given the information in the following sketch, find the force F required to lift the load.

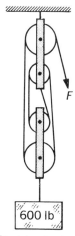

Sketch of setup

Solution

Step 1 Draw an FBD of the lower pulley assembly.

FBD of lower
pulley assembly

Step 2 Because moment summations about each pulley must equal zero, and moment arms for each pulley are equal, the rope forces on each pulley must be equal. As we follow the rope from one pulley to the next and analyze each pulley, we find that all of the rope forces are equal. Therefore, summing the vertical forces on the lower pulley assembly, we have:

$\Sigma F_y = 0 = -600 \text{ lb} + 4F$

$F = 150 \text{ lb}$

NONCONCURRENT FORCE SYSTEMS

Of all the force systems we will discuss, the nonconcurrent force systems may prove to be the most difficult to understand. So plan to spend extra time working the problems. The solutions of these systems are most likely to require all of the statics equilibrium equations:

$\Sigma M = 0$
$\Sigma F_x = 0$
$\Sigma F_y = 0$

Fortunately, the method of solution is similar to that for parallel force systems. The difference is that the forces may be in any direction. You must use your judgment to decide whether or not a force should be broken into its components for easier solution.

Procedure for Solving Nonconcurrent Force Problems

Step 1 Draw an FBD of the structural member involved.

Step 2 Choose a moment center to eliminate one or more unknown forces. Using the summation of moments, determine the magnitude of one unknown force. If a force is at an angle to the X- and Y-axes, you may want to use components.

Step 3 Sum forces in the X and Y directions to solve for other unknown forces.

Step 4 If you obtain components for an answer, you may want to resolve these into a single force using the Parallelogram Law.

SAMPLE PROBLEM 3.5
Nonconcurrent Forces in Equilibrium

Problem

Given the boom and load shown in the following figure, find the reactions at A and B.

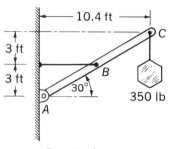

Sketch of setup

Solution

Step 1 Draw an FBD of the boom. We know the directions of the forces at B and C, but not at A. Therefore, the X and Y components of the force at A are shown.

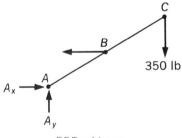

FBD of boom

Step 2 Moments will be summed about A.

$\Sigma M_A = 0 = +(350 \text{ lb} \times 10.4 \text{ ft}) - (B \times 3 \text{ ft})$
$B = 1213 \text{ lb} \leftarrow$

Step 3 Sum forces in the X and Y directions.

$\Sigma F_x = 0 = + A_x - 1213$
$A_x = 1213 \text{ lb} \rightarrow$
$\Sigma F_y = 0 = + A_y - 350$
$A_y = 350 \text{ lb} \uparrow$

Step 4 Resolve the components at A into a single force. A sketch of the parallelogram is helpful.

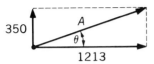

Components of force A

$$\theta = \tan^{-1}\left(\frac{350}{1213}\right) = 16.1°$$

$$A = \frac{350}{\sin 16.1°} = 1260 \text{ lb}$$

Based on the fact that Sample Problem 3.5 has three forces in equilibrium, you are probably thinking that it could be solved as a concurrent force problem, as was done in Sample Problem 2.4. That is true. But, if we consider the components of A to be separate forces, which is usual in more complicated structures, we do have four nonconcurrent forces in equilibrium. Also, you now have two methods of solution for this particular setup.

REPLACING ONE FORCE SYSTEM WITH ANOTHER

There are times when technicians are required to replace one force system acting on a structure with another equivalent force system. This may be necessary before determining the balancing forces required for equilibrium or for determining the internal forces in the structure. Later in this text, you will be solving problems related to curved machine members or structures, eccentrically loaded structures, and coiled springs. In all of these situations, you will need to know how to replace the given load with an equivalent load at another position.

Recall Figure 3.3(b). The forces on the pulley could be moved from the point of concurrence to the hub. Our present task is similar, except that now we have moments to contend with. The basic rules to follow are:

1. The force in the new system must equal the original force or forces in magnitude and direction.

CHAPTER THREE

2. The moment of the new system must equal the original moments (about the same moment center) in magnitude and direction.

Expressed algebraically:

$$\text{new } F_x = \text{original } \Sigma F_x$$
$$\text{new } F_y = \text{original } \Sigma F_y$$
$$\text{new } M = \text{original } \Sigma M$$

Let's start with a situation you will encounter later in this text. Study Sample Problem 3.6. The problem is to replace the given force with a force through O. The original system had a 600 lb·in. moment applied about point O. Since the replacing force would have no moment about O, a moment must be applied. As no other single force can be added, we must end up with a couple to supply the moment about O. Therefore, we replace the original system with a single 200 lb force directed down through O and a 600 lb·in. CW couple applied about O.

SAMPLE PROBLEM 3.6
Replacing a Single Force with a Force and a Couple

Problem

Refer to the following sketch and replace the given force with a force system applied at O.

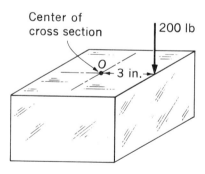

Solution

The present conditions are:

$F_y = -200$ lb
$M_o = +200$ lb (3 in.) $= 600$ lb·in. CW

Placing the 200 lb force at O, we note it has a zero moment arm and therefore does not apply a moment about O. A moment must be applied, and a net force cannot be added. Therefore, a couple must be applied. Exactly *how* the couple is applied is not our concern at this point of a design problem. Only the magnitude and direction of the couple need be specified.

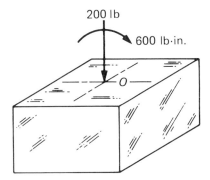

If the original force system consists of a number of forces, then use the following procedure.

Procedure for Replacing a Nonconcurrent Force System

Step 1 Draw an FBD, and choose a moment center if none is specified.
Step 2 Resolve the original forces into a single force as if they were concurrent.
Step 3 Add the moments algebraically to obtain a single moment value.
Step 4 Show the answer with a force through the moment center and a couple; or show the answer with the force offset from the moment center to provide the correct moment.

Sample Problem 3.7 illustrates how several nonconcurrent forces can be replaced with another force system.

SAMPLE PROBLEM 3.7
Replacing a Nonconcurrent Force System

Problem

Replace the given force system with (a) a force and a couple acting at O and with (b) a single force.

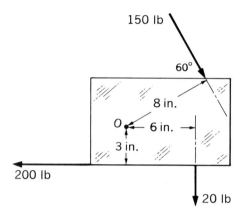

Solution (a)

Step 1 The FBD and moment center have already been given in the figure above.

Step 2 Resolve the original forces into a single force. (The summation of both forces and moments is presented in tabular form below. You can check for missing terms simply by glancing at this information.) The rectangular components of the 150 lb force are calculated first.

$$F_x = +150 \text{ lb} \times \cos 60° = +75 \text{ lb}$$
$$F_y = -150 \text{ lb} \times \sin 60° = -130 \text{ lb}$$

Step 3 The tabular information provides the X and Y components of the single force, and the single moment.

Force →	200 lb	150 lb	20 lb		SF
$\Sigma F_x =$	−200	+75	0	=	−125
$\Sigma F_y =$	0	−130	−20	=	−150
$\Sigma M_o =$	+600	+1200	+120	=	+1920

Step 4 The parallelogram is sketched below, and the magnitude and angle of the single force are calculated.

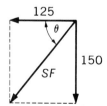

$$\theta = \tan^{-1}\left(\frac{150}{125}\right) = 50.2°$$

$$SF = \frac{150}{\sin 50.2°} = 195 \text{ lb}$$

The complete answer appears in the following figure.

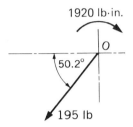

Solution (b)

Step 4 Since a single force is required, its position must be moved so that it has a moment arm sufficient to provide a CW moment of 1920 lb·in.

$$M = F \times D$$
$$1920 \text{ lb·in.} = 195 \text{ lb} \times D$$
$$D = 9.8 \text{ in.}$$

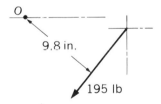

SUMMARY

A moment is equal to the product of a force and a perpendicular distance to the moment center.

$$M = F \times D$$

Varignon's Theorem states that the moment of a force is equal to the sum of the moments of its components.

Parallel force systems require just two equilibrium equations:

$$\Sigma F_y = 0$$
$$\Sigma M = 0$$

The forces on each end of a rope passing around a pulley are equal in magnitude, because the pulley radius is the moment arm for each force. Since moment arms are equal, the forces must be equal to satisfy the equilibrium equation $\Sigma M = 0$.

Nonconcurrent force systems require all three statics equilibrium equations:

$$\Sigma F_x = 0$$
$$\Sigma F_y = 0$$
$$\Sigma M = 0$$

Replacing one force system with another requires that the forces and moments in the replacing system equal the forces and moments in the original system.

PROBLEMS

Moments

1. You are using a torque wrench 18 in. long. Your instructions are to tighten a particular nut with a torque of 75 lb·ft. How much force must you exert?

Parallel Force Systems

2. Refer to Figure 3.4. Determine the resisting forces at A and B.
3. Refer to Figure 3.5. Determine the resisting force system at A.
4. Refer to Figure 3.6. Two knife edges at A restrain the up-and-down movement of the bar; however, only one knife is loaded at a time. Determine the resisting forces at A and B.
5. Refer to Figure 3.7. If the bar is supported on rollers at A and C, determine the resisting forces.
6. Refer to Figure 3.7. If the beam is supported by knife edges at A and B, determine the resisting forces.
7. Refer to Figure 3.8. Determine the resisting forces at A and B.

Nonconcurrent Force Systems   51

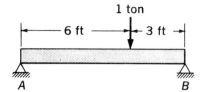

FIGURE 3.4   Problem 2

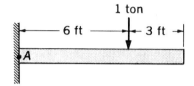

FIGURE 3.5   Problem 3

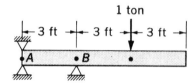

FIGURE 3.6   Problem 4

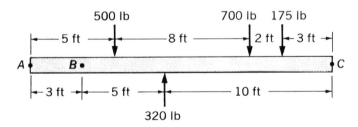

FIGURE 3.7   Problems 5 and 6

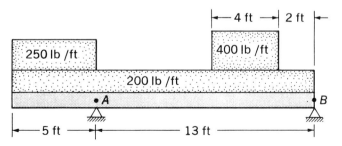

FIGURE 3.8   Problem 7

52 CHAPTER THREE

8. Solve Sample Problem 3.2, but place the resisting forces 5 ft and 15 ft to the right of A.
9. (SI Units) Solve the problem represented in Figure 3.4 if the distances are in metres and the load is 2 MN.

Couples

10. A step pulley has forces applied as shown in Figure 3.9. The 33 lb forces are parallel, as are P_1 and P_2. Determine the forces P_1 and P_2 necessary for equilibrium.
11. Three pulleys A, B, and C are of equal diameter and are mounted along a shaft. A torque of 125 lb·ft CW is applied to A, and a torque of 85 lb·ft CCW is applied to B. Determine the torque on pulley C for equilibrium.

Pulleys

12. Figure 3.10 indicates a rope passing over a pulley and supporting a weight of 70.7 lb. Determine the resisting force at the wall and at sections AA and BB and the reaction of the axle on the pulley.
13. In Figure 3.11(a) and (b), determine the weights marked W needed to balance the systems.
14. Of the two alternatives shown in Figure 3.12(a) and (b), which applies the *least* force to the tree? Assume all lines between pulleys are horizontal.
15. (SI Units) Solve Problem 11; this time the torque on A is 200 N·m CW and on B is 10 kN·m CCW.

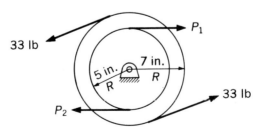

FIGURE 3.9 Problem 10

Nonconcurrent Force Systems 53

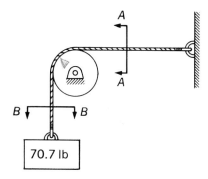

FIGURE 3.10   Problem 12

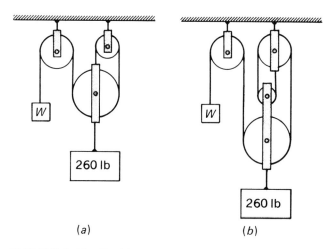

(a)   (b)

FIGURE 3.11   Problem 13

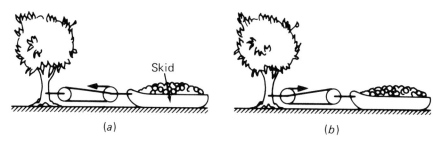

(a)   (b)

FIGURE 3.12   Problem 14

Nonconcurrent Force Systems

16. Refer to Figure 3.13(a) and (b). Solve for the horizontal and vertical resisting forces at A and the vertical resisting force at B.
17. Using the information in Figure 3.14, solve for equilibrium with the necessary forces applied at A and B.
18. Solve Sample Problem 2.4 using moment summations. *Hint:* Start with a moment center at the bottom of the ladder first. Also, when using the top of the ladder as a moment center, work with the X and Y components of the ground force.
19. Solve Sample Problem 2.4, but add 35 lb as the weight of the ladder.
20. Solve for the reactions on the boom in Figure 3.15. The boom is 10 ft long, and the load weighs 200 lb and is placed 5 ft from the wall. The wire stay is fastened to the end of the boom at a 45° angle.
21. Figure 3.16 shows a boom similar to the one in Figure 3.15, except that the 200 lb load is at the end of the boom, and the wire stay is fastened to the midpoint of the beam at a 60° angle. Solve for the reactions on the boom.
22. (Extra Credit) Refer to Figure 3.17. Place the body in equilibrium (a) with two resisting forces, one vertical and the other at an angle of 45° to the horizontal, and (b) with two resisting forces, each 30° to the horizontal.
23. We are given a setup similar to that in Sample Problem 3.6, except that a single upward force of 5000 lb is applied 2.5 in. to the right of O. Replace with a single force through O and a couple about O.

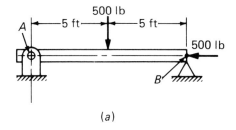
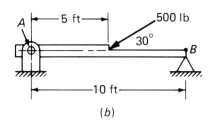

FIGURE 3.13 Problem 16

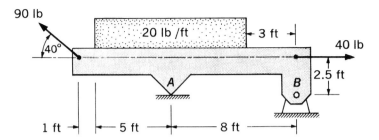

FIGURE 3.14 Problem 17

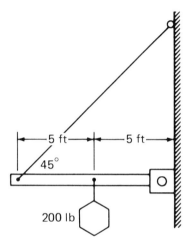

FIGURE 3.15 Problem 20

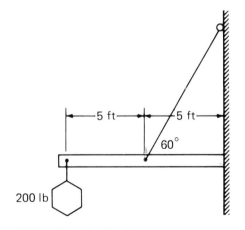

FIGURE 3.16 Problem 21

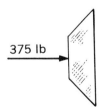

FIGURE 3.17 Problem 22

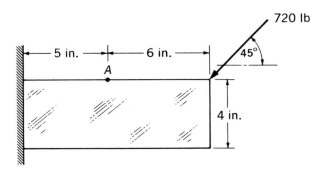

FIGURE 3.18 Problem 24

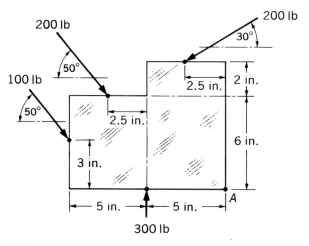

FIGURE 3.19 Problem 25

24. Refer to Figure 3.18. Replace the given force with a single force through A and a couple about A.
25. Refer to Figure 3.19. Replace the given forces with a single force and a couple located at point A.
26. Refer to Figure 3.20. Replace the given forces with a single force and a couple located at point A.
27. (Extra Credit) In Figure 3.21, a vertical force F has a moment of -800 lb·ft about A and a moment of $+1200$ lb·ft about B. Determine the distance X and magnitude of F.
28. (Extra Credit) Given the information in Figure 3.22 and that the horizontal component $F_h = -50$ lb and the moment of the force F about B is $+300$ lb·ft CW, determine (a) the magnitude of F_v, (b) the horizontal intercept of the line-of-action of F, and (c) the magnitude of F.

Nonconcurrent Force Systems 57

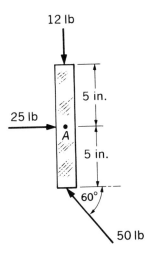

FIGURE 3.20   Problem 26

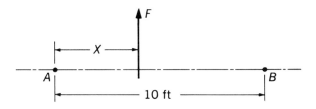

FIGURE 3.21   Problem 27

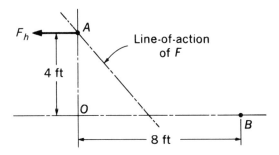

FIGURE 3.22   Problem 28

CASE STUDY

A recent tragedy in the United States involved the collapse of two balconies suspended from a ceiling, one above the other. See Figure 3.23(a). The long suspension rods supported box beams, which in turn supported the balcony flooring. Each box beam was formed by two channels fastened together, as shown in Figure 3.23(b). The disaster occurred when, under the weight of a crowd of people, at least one of the nuts supporting a box beam pulled through the beam. Upon evaluation of the failure, it was discovered that the design was changed during construction from that shown in Figure 3.23(c) to the one shown in Figure 3.23(d). Note that, in the original design, both balconies were attached to the same long suspension members. In the final design, a separate set of suspension members supported the lower balcony. Your job is to decide whether or not the design change may have been a factor in the failure. Give your reasons. Check the load on the suspension members and on each nut for both the original and final designs. Assume that the total weight of people on each balcony was the same.

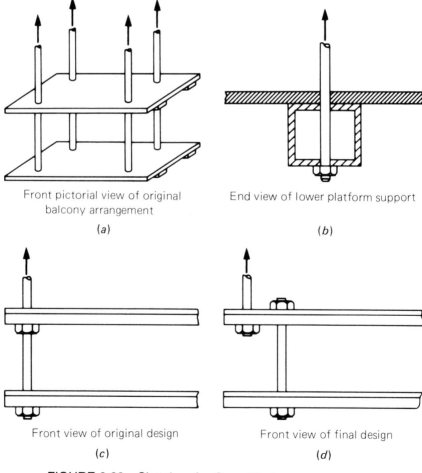

Front pictorial view of original balcony arrangement
(a)

End view of lower platform support
(b)

Front view of original design
(c)

Front view of final design
(d)

FIGURE 3.23 Sketches for Case Study

COMPUTER PROGRAM

The following program will compute the reactions on a simple beam (supports at each end) or on an overhanging beam. The beam can overhang on either end or both ends. Loads must be concentrated. If the problem has a distributed load, calculate the total weight and location of the center of gravity, and put these into the computer.

```
10   REM   BEAMREACTION
20   PRINT "THIS PROGRAM WILL COMPUTE THE REACTIONS ON A SIMPLE BEAM OR"
21   PRINT "OVERHANGING BEAM WITH ANY NUMBER OF CONCENTRATED LOADS."
22   PRINT "LEFT REACTION=A, RIGHT REACTION=B. DISTANCES IN FEET, LOADS IN
        POUNDS.
23   PRINT
24   PRINT "TYPE IN THE NUMBER OF LOADS."
30   INPUT N
35   PRINT
40   PRINT "TYPE IN THE AMOUNT OF EACH LOAD, IN SEQUENCE, STARTING AT BEAM
        'S LEFT END."
50   FOR F = 1 TO N
60   INPUT M(F)
70   NEXT F
80   REM   --- "LD" IS THE TOTAL LOAD
90   LET LD = 0
100    FOR F = 1 TO N
110    LET LD = LD + M(F)
120    NEXT F
130    PRINT
140    PRINT "TYPE IN THE DISTANCE OF EACH LOAD, IN SEQUENCE, STARTING"
141    PRINT "FROM THE LEFT END OF THE BEAM."
150    FOR D = 1 TO N
160    INPUT A(D)
170    NEXT D
180    PRINT
190    PRINT "TYPE IN THE DISTANCE FROM THE LEFT END OF A, THEN B."
200    INPUT DA,DB
210    LET Z = 0
220    FOR I = 1 TO N
230    LET M = M(I) * (A(I) - DA)
240    LET Z = Z + M
250    NEXT I
260    REM   ---"Z" IS THE SUM OF THE MOMENTS ABOUT A.
270    LET B = Z / (DB - DA)
280    LET A = LD - B
290    PRINT "TOTAL LOAD = ";LD;" LB."
300    PRINT
310    PRINT "REACTION AT A = ";A;" LB."
320    PRINT
330    PRINT "REACTION AT B = ";B;" LB."
340    END
```

Framed Structures and Friction

OBJECTIVES

The previous chapters introduced you to equilibrium and force systems. This chapter applies that knowledge to examine framed structures and friction. Upon completing this chapter, you will be able to (1) perform the preliminary analysis of forces in the members of trusses and other frames and (2) apply properly the friction formula: $F_f = \mu \cdot N$.

INTRODUCTION

Framed structures are structures composed of individual parts or members connected to each other. Our interest lies in those structures designed to support loads, like bridges. To illustrate how seriously bridge building has always been considered, even in early times, let us consult history. In 480 B.C., for example, when a partially completed Persian army pontoon bridge across the Hellespont was destroyed by a storm, the chief engineer was beheaded and the water given hundreds of lashes for punishment. Other records indicate that the Romans threw virgins into the waters as a sacrifice to the river gods whenever important bridges were completed.

Trusses are crucial in bridge construction. They were used in bridge design as early as the 1500s, but there was no clear understanding of the forces acting on the truss members until the early 1800s. To begin our understanding, we must start with *two-force* members.

TWO-FORCE MEMBERS

A two-force member is held in equilibrium by just two equal, opposite, and collinear forces. A rope or wire, as in Figure 4.1(a), is a two-force member that can resist tensile forces only. If a body is more rigid than a rope and can resist tensile and compressive forces in many directions,

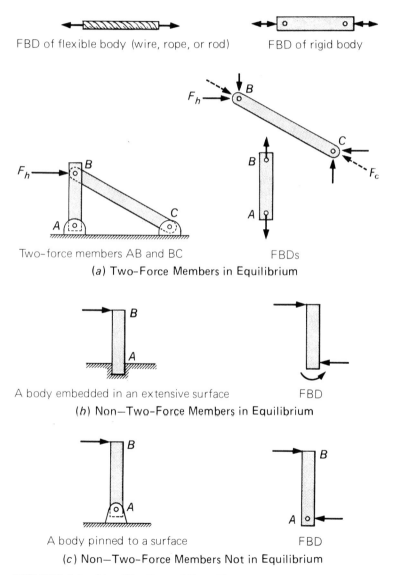

FIGURE 4.1 Identification of Two-Force Members

how can we be sure it is behaving like a two-force member? The sketches in Figure 4.1 show why a two-force member must be connected at its ends by frictionless pins and why no forces must be applied to the member except by the pins. Figure 4.1(b) shows a member that is in equilibrium, but it is not a two-force member. In Figure 4.1(c), the member is pinned at A, but the force at B creates a moment about A that the pin cannot resist, and the member will fall over. Note that in Figure 4.1(b) and (c), the member AB is not considered a two-force member. Also, in Figure 4.1(c), the member is not in equilibrium. However, let us attach a member BC by a pin at B, as shown in Figure 4.1(a). Again we have a horizontal force applied at B. It appears logical that the structure is capable of resisting the force, yet the force is not applied along the axis of either member, and its line-of-action does not pass through A or C. Drawing free body diagrams of both members, we note that AB cannot resist any horizontal force; therefore, member BC must resist the entire horizontal force. The resisting force at C in Figure 4.1(a) has a horizontal component that is equal and opposite to the applied horizontal force F_h. Knowing the angle of member BC, we can solve for the vertical component of the force at C as well as the main force along the longitudinal axis of the member. Moving up to the B end of member BC, the vertical component of the force on BC must be equal and opposite to the vertical component at C. The horizontal component is F_h. Our analysis now shows member BC in equilibrium. The next question is, "What supplies the downward component at B acting on member BC?" The horizontal force cannot. Therefore, member AB must. Since AB is pulling down on BC, then BC must be pulling *up* on member AB as is shown in the free body diagram in Figure 4.1(a).

TRUSS STRUCTURES

A *truss* is a framed structure composed of a number of members, all of which are *two-force members*. In addition, the members of a truss must form triangles, or the structure may collapse. See Figure 4.2 for an illustration. You can see from Figure 4.2(b) that the structure can assume various positions under different force systems. To ensure that a structure is a truss, the following conditions must exist:

1. The members must be loaded only at their ends.
2. The members must be connected to each other and receive loads

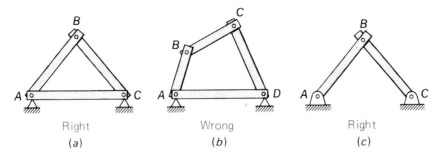

FIGURE 4.2 Truss Structures

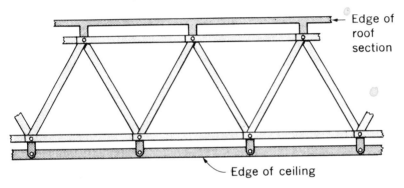

FIGURE 4.3 Section of Roof Truss

through pins* (assumed frictionless) at the ends of the members.
3. The weight of the member is assumed negligible compared to the forces it must resist.

Figure 4.3 shows how a roof truss might support its loads. Note that the loads are applied at the joints (pins). Figure 4.4 pictures a number of standard types of trusses. The Howe and Pratt trusses shown were originally designed as through-bridge trusses (loads were applied at the bottom joints). The Howe truss has its vertical members in tension. The Pratt truss has its diagonal members in tension. Orig-

---

*You may observe that the members of some trusses are riveted or welded at the connections. These may still be considered pinned connections in the first stages of design.

64　CHAPTER FOUR

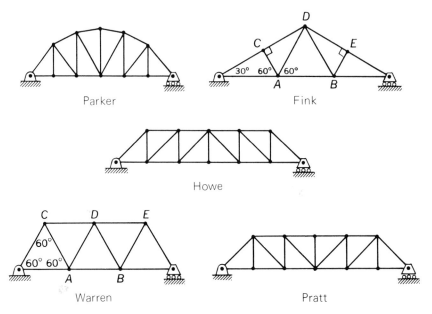

FIGURE 4.4　Types of Trusses

PHOTO 4.1　Truss Bridge across the Rappahannock River

inally, the tension members were iron rods, and the compression members were timber. The Fink truss is a roof truss.

There are several methods for solving truss problems, but this text will consider only the procedure called the *method of joints*. Figure 4.5 illustrates the steps in this procedure, starting with the sketch in Figure 4.5(*a*).

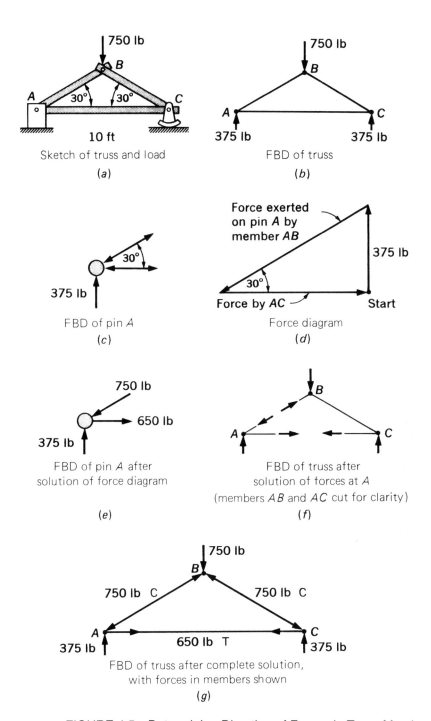

FIGURE 4.5 Determining Direction of Forces in Truss Members

Procedure for the Method of Joints

Step 1 Generally, the first step is to draw a free body diagram of the complete structure and solve for the reactions at the truss supports. See Figure 4.5(b).

Step 2 A free body diagram is made of each joint (pin), showing the lines-of-action of the forces acting on the pin and the directions of the forces that are known.

Step 3 Starting at the pin with the least unknown forces and working in sequence across the truss, draw the force diagram and solve for the unknown forces. See Figure 4.5(c) and (d).

Step 4 Arrows are placed on the force triangle in Figure 4.5(d), indicating the direction of each force; this information is related back to the free body diagram in Figure 4.5(e), and the arrows are placed in their respective positions.

Step 5 If a force is directed toward the pin (squeezing the pin), the force is compressive and the member supplying the force is in compression. If the force is directed away from the pin, it is a tensile force and the member supplying the force is in tension. *Remember:* If a member is in tension, it is exerting a tensile force on *both* of its connecting pins; and if in compression, it is exerting a compressive force on both of its pins.

Step 6 Summarize your answers by listing each member and its force, or indicate the type of force exerted by each member on a free body diagram of the whole truss. Use arrows near the pin, or the letters C or T next to the member, as shown in Figure 4.5(g).

ZERO-FORCE MEMBERS

Frequently, members are placed in a truss to maintain the rigidity of the truss, not to take a load. Or, a member in a roof truss may be designed to resist, say, a snow load but not a ceiling load. An example is shown in Figure 4.6. Member BD cannot exert a force. For the reason, look at a free body diagram of pin D. Assume that BD is exerting a compressive force on pin D, and members AD and CD exert tensile forces on pin D. Note that AD and CD oppose each other, but there is no force to oppose BD. Therefore, equilibrium cannot be maintained if a force is applied by BD to pin D. This is true if BD is at any angle to the horizontal in Figure 4.6. A zero-force member can be identified by

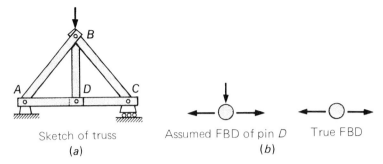

FIGURE 4.6 Illustration of a Zero-Force Member

inspection from the following statement:

If three members of a truss meet at a joint that has no external force applied, and two of the members are collinear, then the third member can have no force exerted on or by it.

If some external vertical load were applied at pin D in Figure 4.6, then it follows from what has been said above that member BD must support the entire vertical applied force. Sample Problems 4.1 and 4.2 solve truss problems.

SAMPLE PROBLEM 4.1
Cantilever Truss

Problem

Given the information in the following sketch, find the force in each member of the truss and indicate whether the force is tensile or compressive. Note that the term *kip* represents kilopounds. One kip is 1000 pounds.

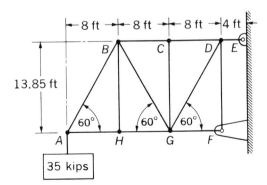

Solution

Determine the reactions at F and E. (The procedure previously listed will be followed, but the steps will not be indicated, because Steps 2, 3, and 4 are repeated for each joint or pin.) By inspection, E cannot take a vertical force. Therefore:

$F_v = 35$ kips ↑

$\Sigma M_F = 0 = -35$ kips $(24$ ft$) + E_h (13.85$ ft$)$

$E_h = 60.6$ kips

Σ horizontal forces $= 0 = +60.6$ kips $- F_h$

$F_h = 60.6$ kips ←

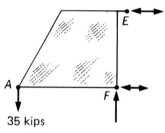

FBD of complete truss

Pin A Start with pin A and work around the truss.

$AB = \dfrac{35}{\sin 60°} = 40.4$ kips T

$AH = 35 (\cot 60°) = 20.2$ kips C

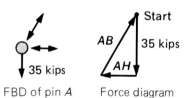

FBD of pin A Force diagram

Pin B By inspection, $BH = 0$. This results in an equilateral triangle for the force triangle. Therefore:

$AB = BG = BC$

$BG = 40.4$ kips C

$BC = 40.4$ kips T

Framed Structures and Friction   69

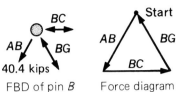

FBD of pin B   Force diagram

Pin C By inspection, $CG = 0$. Therefore, $CD = 40.4$ kips T.

$CD = 40.4$ kips T
$CG = 0$

FBD of pin C

Pin D E_h has already been solved, and

$DE = E_h = 60.6$ kips T

$$DF = \frac{60.6 - 40.4}{\cot 60°} = \frac{20.2}{\cot 60°} = 35 \text{ kips C}$$

$$DG = \frac{20.2}{\cos 60°} = 40.4 \text{ kips T}$$

FBD of pin D   Force diagram

Pin F By inspection,

$FG = 60.6$ kips C

FBD of pin F

Pin G Since DG and BG form two legs of an equilateral triangle,

$GH = FG - 40.4 = 60.6 - 40.4 = 20.2$ kips C

CHAPTER FOUR

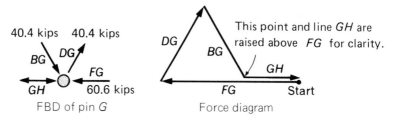

FBD of pin G   Force diagram

Pin H This is a check. From the FBD of pin H, we can see that both AH and GH should have the same values, and they do.

FBD of pin H

A summary of the members and their forces is given in the accompanying table.

Reaction at E Horizontal Only	60.6 kips →
Reaction at F Horizontal Component Vertical Component	60.6 kips ← 35 kips ↑
Member	Forces in kips
AB AH BC BG BH CD CG DE DF DG FG GH	40.4 T 20.2 C 40.4 T 40.4 C 0 40.4 T 0 60.6 T 35.0 C 40.4 T 60.6 C 20.2 C

METRIC SAMPLE PROBLEM 4.2
Cantilever Truss
Problem

Solve for the forces on each member. *Note:* Loads are frequently given in mass units (kg) and technically should be changed to force units (N) for solution.

Framed Structures and Friction   71

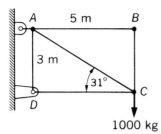

Solution

Change to force units:

1000 kg = 1000 (9.8) = 9800 N

Determine the reaction at A. There is no vertical component. Therefore:

$\Sigma M_D = 0 = 9800\ (5\ m) - A_x\ (3\ m)$
$A_x = 16\ 300\ N \leftarrow$

Determine the reaction at D. By inspection,

$D_y = 9800\ N \uparrow$

The horizontal reaction at D has to oppose the reaction at A. Therefore:

$D_x = 16\ 300\ N \rightarrow$

There are too many unknowns to start at A. Therefore, start at pin D.

FBD of pin D

Pin D   There is no need to draw a force diagram for pin D, because the forces on AD and CD can be determined by inspection.

$AD = 9800\ N\ (C)$
$CD = 16\ 300\ N\ (C)$

Pin B   By inspection, we can see that any force AB that might apply to the pin cannot be opposed. The same is true for member BC. Therefore:

$AB = 0$
$BC = 0$

72 CHAPTER FOUR

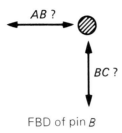

FBD of pin B

Pin C The FBD aids us in constructing the force diagram.

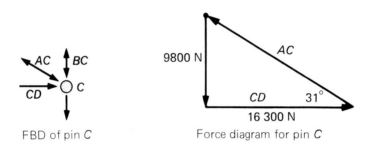

FBD of pin C   Force diagram for pin C

$$AC = \frac{16\,300}{\cos 31°} = 19\,000 \text{ N (T)}$$

A summary of the members and their forces is given in the accompanying table.

Reaction at A_x	16 300 N ←
Reaction at D_x Reaction at D_y	16 300 N → 9 800 N ↑
Member	Forces in Newtons
AB AC AD BC CD	0 19 000 T 9 800 C 0 16 300 C

OTHER FRAMED STRUCTURES

Although trusses come under the heading of framed structures, there are other types of framed structures supporting loads. The type investigated here will be referred to simply as *frames* to differentiate them from trusses.

The frames discussed here have members that support loads at locations between their end points. The loads may be at any angle to the longitudinal axis of the member. A frame may also include two-force members, such as member *AB* of the frame analyzed in Sample Problem 4.3

You have already solved ladder problems. A ladder leaning against a wall can be classified as a frame because some of the loads are applied between the ends of the ladder. Sample Problem 4.3 considers a more involved frame. The following procedure is helpful in solving frames.

Procedure for Solving Frame Problems

Step 1 Draw a free body diagram of the whole structure and attempt to solve for the reactions at the structure supports. The word *attempt* is used because if the structure has more than two supports (see Figure 4.23), you may have to start with one member independently and work toward solution of the other members.

Step 2 Draw a free body diagram of each member, putting in the vertical and horizontal components of forces at each connecting pin.

Step 3 Determine whether or not there are any two-force members in the frame in order to facilitate solution.

Step 4 Choose a member whereby a moment equation can be written with only one unknown.

Step 5 It may be necessary to shift solution to another member before completing solution of the first.

Step 6 Note that at a pin connection, the force applied to one member is equal, opposite, and collinear to the force applied to the connected member, as shown on each member's free body diagram.

SAMPLE PROBLEM 4.3

Frame Structure

Problem

Given the A-frame and loads in the following sketch, determine the forces on each member of the frame.

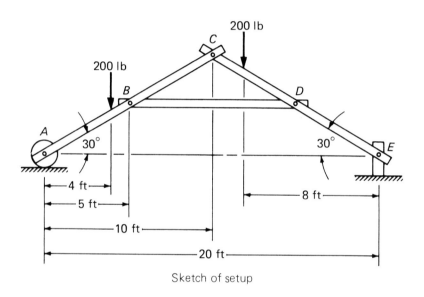

Sketch of setup

Solution

First, draw the FBD of the complete frame and find the reactions at A and E:

$\Sigma M_A = 0 = +(200 \text{ lb} \times 4 \text{ ft}) + (200 \text{ lb} \times 12 \text{ ft}) - (E \times 20 \text{ ft})$
$E = 160 \text{ lb} \uparrow$

Note: There are no horizontal forces applied to the frame, so there are no horizontal reactions at E.

$\Sigma M_E = 0 = -(200 \times 8) - (200 \times 16) + (A \times 20)$
$A = 240 \text{ lb} \uparrow$
Check: $\Sigma F_y = 0 = -200 - 200 + 240 + 160$

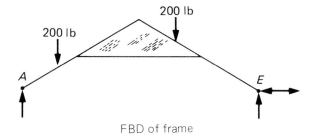

FBD of frame

Second, draw the FBD of member ABC and solve for the forces acting on it.

$\Sigma M_C = 0 = +(240 \times 10) - (200 \times 6) - (B \times 2.89)$
$B = 830 \text{ lb} \rightarrow$

The easiest way to find the forces at C is to sum horizontal and vertical forces on ABC.

$\Sigma Y = 0 = +240 - 200 - C_y$
$C_y = 40 \text{ lb} \downarrow$
$\Sigma X = 0 = +830 - C_x$
$C_x = 830 \text{ lb} \leftarrow$

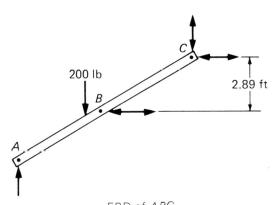

FBD of ABC

Third, draw the FBD of member CDE and solve for the forces acting on it. Member BD is in tension; therefore, it is pulling on both ABC and CDE. At C, member ABC has a horizontal compressive force applied; therefore, ABC must be *pushing* CDE horizontally. At C, member CDE is pushing *down* on ABC; therefore, ABC must be pushing *up* on CDE.

Check: $\Sigma Y = 0 = +160 + 40 - 200$
Check: $\Sigma X = 0 = +830 - 830$

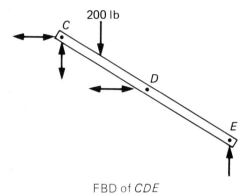

FBD of CDE

Answer

The answers are given on the FBD of each member.

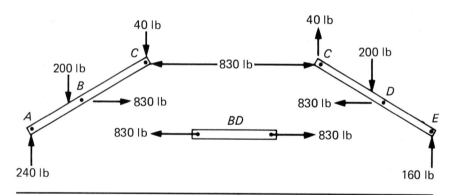

STATIC FRICTION

Picture a heavy metal block on an extensive smooth dry surface, with a horizontal force applied. The force is gradually increased from zero up to some maximum force necessary just to start motion of the block. Once motion has started, the force necessary to maintain motion is reduced. Frictional resistance necessarily opposes the direction of motion or impending motion. Prior to motion of the block, we have a static frictional resistance; after motion begins, we have a sliding, or kinetic, frictional resistance. This text is primarily concerned with the *maxi-*

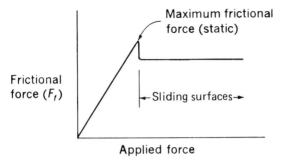

FIGURE 4.7 Graph of Frictional Force *vs.* Applied Force

mum static frictional resistance to an applied force, in other words, *the maximum force that opposes impending motion*. Figure 4.7 graphically illustrates that the frictional resisting force is reduced once motion is started.

The causes of friction are still being investigated, because there always seems to be some exception to each theory presented. It used to be thought the major cause of friction (the resistance to sliding) was surface roughness—even a surface that appears smooth to the eye and to the touch actually has high and sharp protrusions and deep valleys on a microscopic scale. The explanation for static frictional resistance was that to start motion, the hills of one surface had to be pushed up and out of the valleys of the other surface. Once motion was started, the protrusions bumped over each other without falling into the valleys; thus, sliding friction was less than static friction. Surface roughness is still a factor but is no longer considered the primary reason for friction, because surfaces that are made as smooth as obtainable today begin to show an increase in frictional resistance. Another factor is that sharp protrusions on one surface can actually puncture the other contacting surface and create frictional resistance.

It is now believed that frictional forces are primarily produced by the interaction between surface atoms. In other words, when two surfaces are placed together, a very few tiny areas are in close enough contact to actually weld together. When sliding occurs, there apparently is insufficient time for strong atomic bonds to develop; therefore, sliding resistance is lower than static resistance.

An interesting phenomenon occurs in space and in a high vacuum. When rubbed, some metals actually weld together in a strong bond. Scientists believe that in ordinary situations, every surface has an extremely thin film of oxide, moisture, and microscopic dirt and grease particles that help lubricate the surface. In a vacuum, this film evaporates and gets cleaned away easily with rubbing, allowing large areas of the surfaces to come close enough for more atoms to bond together.

Coefficient of Friction

It has been found experimentally that the *maximum* static frictional resistance varies directly as the *normal*, or perpendicular, reaction force exerted by the surface on the block (or object). This relationship is constant for a given surface condition, and the constant is called the *coefficient of friction*. The coefficients for various materials are listed in handbooks; a few are given in Table 4.1. The symbol is the Greek letter *mu*, μ. The formula is:

$$F_f = \mu \cdot N$$

where

F_f = maximum static frictional resisting force
μ = static coefficient of friction
N = reaction force applied by the surface and normal (perpendicular) to the surface

The friction formula may be used for sliding friction situations with the application of the appropriate sliding friction coefficient. Also, note that the formula does *not* include the area of the contact surfaces! This means that the amount of area in contact between two materials is unimportant and has no effect on the solution of a problem.

You probably are already thinking that car tires are usually deflated for better travel on beach sand. Such deflation increases the tire area in contact with the sand and gives better traction. But that is not a contradiction of the formula. Beach sand will give way if too

TABLE 4.1 Coefficients of Friction

Surfaces in Contact	Static Friction Coefficient
Mild steel on mild steel	0.74
Wood on metal	0.2–0.6
Aluminum on aluminum	1.05
Glass on glass	0.94
Rubber tire on dry asphalt	0.8
Teflon on teflon or steel	0.04
Steel on ice	0.014 (sliding)

Note: The values were obtained experimentally for clean, dry surfaces.

Angle of Friction

The forces N and F_f may be considered components of a single force applied at an angle theta (θ), measured from N, as shown in sketch (c) of Sample Problem 4.4. The tangent of angle theta is:

$$\tan \theta = \frac{F_f}{N}$$

But, from the friction formula,

$$\mu = \frac{F_f}{N}$$

Substituting equals for equals, we can see that

$$\mu = \tan \theta$$

The angle theta is commonly referred to as the *angle of friction*.

Procedure for Solving Friction Problems

Step 1 Sketch the setup and an FBD of the object in question. Determine the direction of the frictional force (F_f), and place this on the FBD. *Note: F_f opposes impending motion.*

Step 2 Use either of two approaches.
   (a) Resolve F_f and N into a single force SF, and employ the force triangle to solve for the unknowns.
   (b) Sum X and Y components, but place the X-axis parallel to the frictional surface (if it is inclined), and place the Y-axis normal to the frictional surface.

Sample Problems 4.4, 4.5, and 4.6 illustrate frictional problem solutions.

SAMPLE PROBLEM 4.4
Static Friction

Problem

Given the following sketch, and that motion is impending, solve for the coefficient of friction, angle of friction, and reaction SF.

80   CHAPTER FOUR

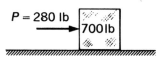

Force required just
prior to motion

Solution

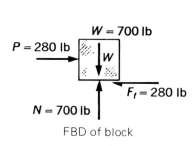

FBD of block

N and F_f replaced
with a single force

$$\mu = \tan\theta = \frac{F_f}{N} = \frac{280}{700} = 0.4$$

θ = angle of friction = 21.7°

$$SF = \frac{F_f}{\sin\theta} = \frac{280}{\sin 21.7°} = 757 \text{ lb}$$

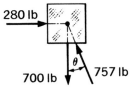

FBD of block with
forces concurrent

Note: The couple formed by P and F_r is analyzed at a later stage of design. For the present, we shall consider all forces in friction problems concurrent at the center of gravity of the block.

SAMPLE PROBLEM 4.5
Static Friction

Problem

Given a 1500 lb block on a 20° incline, and a coefficient of friction (μ) = 0.84, determine (a) whether the block will slide down the incline, (b) what force P, parallel to the incline, is required for impending motion *up* the incline; and (c) what force P is required for impending motion *down* the incline.

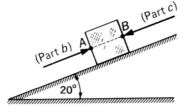

Sketch of setup

Solution (a)

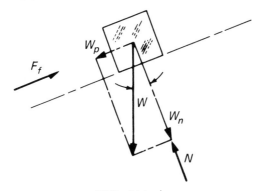

FBD of block

The angle between the weight of the object and N equals 20°. The component of the weight parallel to the incline is:

$W_p = 1500 \sin 20° = 513$ lb

The component of the weight normal to the incline is:

$W_n = 1500 \cos 20° = 1410$ lb $(= N)$

F_f opposing motion down the plane is:

$F_f = \mu N = 0.84 \times 1410 = 1184$ lb

Answer (a)

F_f is greater than the opposing force W_p; therefore, the block will *not* slide down the incline.

Solution (b)

Part (b) can be solved in a manner similar to the solution for part (c), but the force triangle method will be used. The angle of friction (θ) is:

θ = tan⁻¹ 0.84 = 40°

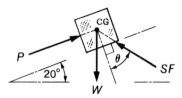

FBD and angle of friction (θ)

The single force SF is in the position shown in the FBD because its component (F_f) must oppose the force P. The angle SF makes with the horizontal (30°) is obtained from the geometry of the setup.

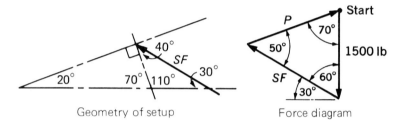

Geometry of setup   Force diagram

Answer (b)

Sine law: $\dfrac{P}{\sin 60°} = \dfrac{1500 \text{ lb}}{\sin 50°}$

P = 1695 lb

Solution (c)

The angle of friction was obtained in part (b) and will be used here.

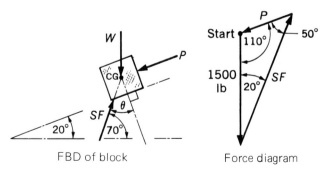

FBD of block   Force diagram

Note: SF is in the direction to resist impending motion.

Answer (c)

Sine law: $\dfrac{P}{\sin 20°} = \dfrac{1500 \text{ lb}}{\sin 50°}$

$P = 670$ lb

SAMPLE PROBLEM 4.6
Friction

Problem

Block A weighs 475 lb. The angle of friction for all surfaces is 20°. A and B are connected by a rope. What must block B weigh to prevent A from sliding down the incline?

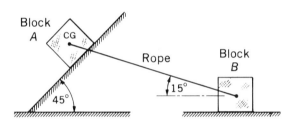

Solution

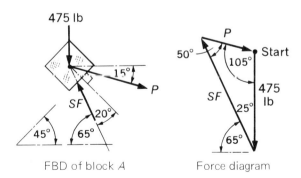

FBD of block A   Force diagram

Sine law: $\dfrac{P}{\sin 25°} = \dfrac{475 \text{ lb}}{\sin 50°}$

$P = 262$ lb tension

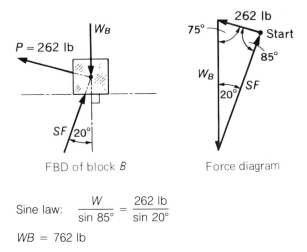

Sine law: $\dfrac{W}{\sin 85°} = \dfrac{262 \text{ lb}}{\sin 20°}$

$W_B = 762$ lb

WEDGES

Wedge problems are handled in a manner similar to the frictional block problems already discussed. However, wedges are probably more confusing because (1) there is a tendency to view the whole structure of wedges and blocks as solidly connected (as in a truss or frame), and (2) the direction of frictional resistance is more difficult to visualize. For these reasons, it is best to draw a free body diagram of each wedge and block and start with the body having the least unknown applied forces. Figure 4.8 illustrates the steps required in solving a wedge type problem. In order to simplify the initial explanation, bar K is a weightless flat bar, and block M is restrained from moving to the right by a wall.

The sketch of the setup in Figure 4.8(a) shows that bar K is to be pushed to the right. Starting with a free body diagram of block M in Figure 4.8(b), we note that the wall force is directed in a horizontal direction (normal to the wall), since block M does not have any tendency to slide up or down the wall. Thus, there is no frictional force. The next step is to decide whether the frictional force F_f on the bottom of the block is directed left or right. This can be confusing, because block M itself doesn't move. Therefore, we must consider the relative motion between K and M. M would tend to move to the left end of bar K; the frictional force acting on M opposes this movement and therefore must be directed to the right, as shown in Figure 4.8(b). With the direction of the frictional force settled, and knowing the angle of friction θ, a

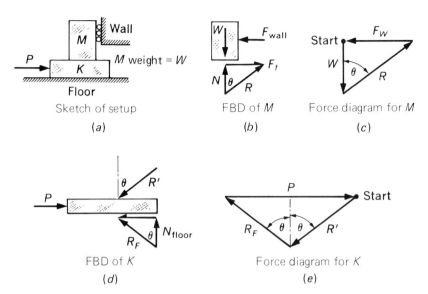

FIGURE 4.8 Steps in Solving a Wedge-Type Problem

single force R can be determined and then used to complete the force diagram in Figure 4.8(c). The use of force triangles will generally be more convenient than resolving forces into X and Y components.

Using the information obtained thus far, we now construct a free body diagram of bar K. Note carefully that the force R was up and to the right, acting on block M, and that the reaction of M on K must be equal, opposite, and collinear. This fact is indicated in Figure 4.8(d) by showing force R' acting down and to the left. Figure 4.8(e) shows the final step in solving for force P. One more reminder: In Figure 4.8(b), N is equal and opposite to W, and in Figure 4.8(d), N_{floor} is also equal to W because the bar K is weightless.

The same procedure illustrated in Figure 4.8 will be used to handle a more difficult problem. The weightless bar K will be replaced by a weightless wedge K, and the wall will offer frictional resistance. Figure 4.9 shows the steps in solution and indicates the directions of forces that can be expected.

In Figure 4.9(a), the wedge has to push against the weight of M as well as against the frictional forces developed at the wall and on two surfaces of the wedge K. Figure 4.9(b) shows that the frictional force of the wall tends to prevent the upward movement of M, and the frictional force of the contacting surface between M and K tends to prevent the block from sliding to the left-hand end of the wedge. The force normal to the wall and the wall frictional force are resolved into

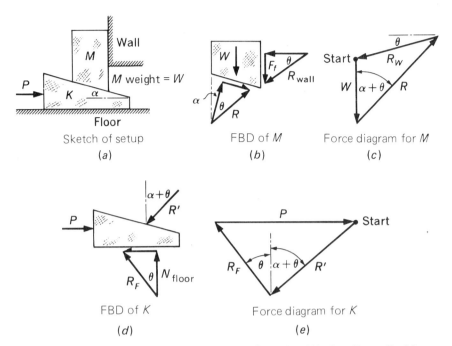

FIGURE 4.9 Steps in Solving a Complex Wedge-Type Problem

a single force R_{wall}. The force applied by wedge K to block M that is normal to the contact surface and the frictional force are combined into a single force R. In order to determine the magnitude and direction of R, the force diagram shown in Figure 4.9(c) is solved. We know that each action has an equal and opposite reaction. Therefore, if the wedge is pushing up on block M with a force R, then M must be pushing down on the wedge with an equal and opposite force R'. This is shown in Figure 4.9(d).

We now continue the solution and direct our attention to the wedge. The floor opposes the wedge motion to the right and therefore supplies a frictional force directed to the left. The force diagram in Figure 4.9(e) can now be constructed. We know the angle and magnitude of R', the angle θ of R_F (but not its magnitude), and that P is horizontal. Using trigonometry, we can solve for P. One word of caution: you should be aware that the value for N in Figure 4.9(d) is *not* equal to the weight of M, but is influenced by the frictional forces acting on the various contacting surfaces. Sample Problem 4.7 solves a wedge problem similar to the one illustrated in Figure 4.9, except that the wedge also will have weight.

Procedure for Solving Wedge Problems

Step 1 Calculate the angle of friction, if necessary. Analyze the positions of the reaction forces with respect to the normals to the contacting surfaces. Then draw the FBDs of the wedge and block.

Step 2 Determine all of the required angles that the forces make with a horizontal line. Then draw the force diagrams.

Step 3 Use the force diagrams to calculate the required forces.

SAMPLE PROBLEM 4.7
Wedges

Problem

As shown in the following sketch, force P is applied to wedge A to lower block B. The coefficient of friction (μ) is 0.306. What is the necessary force P just to start motion of the block?

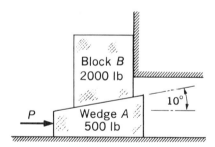

Solution

Step 1 Find the angle of friction (θ) and draw FBDs of A and B.

$\theta = \tan^{-1} 0.306 = 17°$

With reference to forces acting on B: The wall friction resists the downward motion of B; therefore, F_{wall} is upward. If we picture the relative motion between A and B, B slides left over A; therefore, F_A is directed to the right when compared to a line normal to the contact surface. With reference to FBD of A: F_B is equal, opposite, and collinear to force F_A.

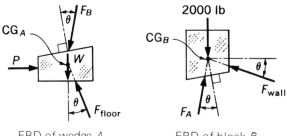

FBD of wedge A   FBD of block B

Step 2 The surface between wedge A and block B is 10° above the horizontal; therefore, the normal to the surface is 110° to the horizontal. F_A (or F_B) is 17° to the normal and therefore 83° to the horizontal. Refer to the sketch that follows.

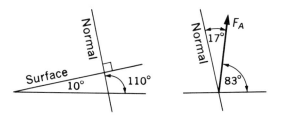

Step 3 The next step is to determine the magnitude of F_A by drawing a force triangle for the forces acting on B. There is no need to calculate the magnitude of the wall force (F_{wall}). Since F_B is the opposite of F_A, we use this information to help complete the force polygon of the forces acting on wedge A and then solve for force P.

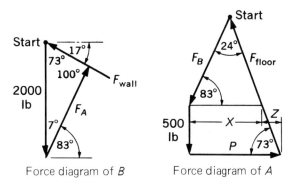

Force diagram of B   Force diagram of A

Sine law: $\dfrac{F_A}{\sin 73°} = \dfrac{2000 \text{ lb}}{\sin 100°}$

$F_A = F_B = 1940$ lb

$P = X + Z$

Sine law: $\dfrac{X}{\sin 24°} = \dfrac{1940 \text{ lb}}{\sin 73°}$

$X = 825$ lb

$Z = 500 (\tan 17°) = 153$ lb

$P = 978$ lb

Two-force members are structural members held in equilibrium by two equal, opposite, and collinear forces, applied along the longitudinal axis of the member. Ropes supporting loads are obviously two-force members in tension.

The ideal truss is a framed structure composed of two-force members connected to each other at their ends by pins that are free to rotate in their fittings. Loads or forces are applied only at the pins.

Zero-force members are placed in a truss to maintain the rigidity of the truss, but they have no forces applied to them.

Other framed structures discussed here generally have more than two forces applied to each member, and these forces need not be applied exclusively to the ends of the members. The major difference in solution of trusses and other framed structures is that for trusses free body diagrams may be drawn of the connecting pins, whereas in other framed structures free body diagrams are drawn of each member.

The maximum static frictional resisting force between two surfaces can be calculated with the formula:

$F_f = \mu \cdot N$

Wedge problems are static friction problems, but their solution requires the analysis of several related contact surfaces.

Two-Force Members and Truss Structures

Draw a free body diagram for each problem, and then solve. Identify forces applied to members of a truss as tensile or compressive. *Note:* 1 kip equals 1000 lb.

1. Two bars are assembled with frictionless pins as shown in Figure 4.10 and are supported at A and C. Which are the correct resisting forces: P_1 and P_2, or F_1 and F_2? Why?
2. Refer to Figure 4.11.
   (a) The 150 lb roller is supported by a rope parallel to the incline. What is the force in the rope? Specify tensile or compressive.
   (b) The 150 lb roller is connected through its axle to a bar pinned to the floor. What force does the bar apply to the roller? Specify tensile or compressive.
3. Solve for the force applied to each member of the structure shown in Figure 4.12.
4. Solve for the force applied to each member of the structure shown in Figure 4.13.
5. Solve for the force applied to each member of the structure shown in Figure 4.14.
6. Refer back to Figure 4.1(d). Let F_h equal 250 lb and the angle at C equal 30°. Solve for the forces acting on members AB and BC.
7. Refer back to Figure 4.5, change the given angles to 50°, and solve.

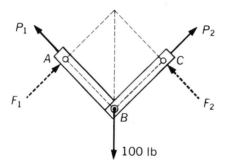

FIGURE 4.10   Problem 1

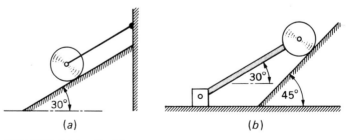

FIGURE 4.11   Problem 2

Framed Structures and Friction   91

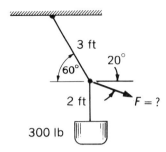

FIGURE 4.12   Problem 3

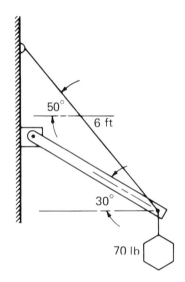

FIGURE 4.13   Problem 4

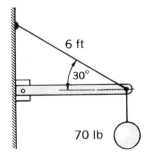

FIGURE 4.14   Problem 5

92 CHAPTER FOUR

8. Refer back to the Fink truss in Figure 4.4. Apply a 1.5 kip downward load to each joint at *A* and *B*. The three lower chord members (the horizontal members) are each 4 ft long. Solve for the force on each member of the truss. Would the force values change if the three lower chord members were each 20 ft in length (angles remain the same)?
9. Refer back to the Warren truss in Figure 4.4. Place a 1.5 kip downward load on each joint *A* and *B*. The three lower chord members are each 10 ft long. Solve for the forces on the truss members.
10. Refer back to the Pratt truss and the Howe truss in Figure 4.4. Place a 100 lb load on each of the joints in the lower chords, except at the supports. Solve each truss and compare. Consider all of the angles to be either right angles or 45° angles, as indicated in the figure.
11. Refer to Figure 4.15. Solve for the force applied to each member of the trusses.
12. (Extra Credit) Refer to Figure 4.16. Solve for the force applied to each member of the trusses.
13. (SI Units) Refer to Figure 4.17. Solve for the force in each member of the truss.

Frame Structures

14. Solve Sample Problem 4.3, but move the frame tie member *BD* down so that it connects with *ABC* just 2 ft from *A*, and with *CDE* 2 ft from *E* (measured horizontally).

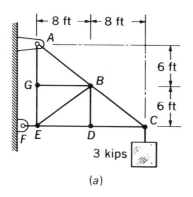
(a)

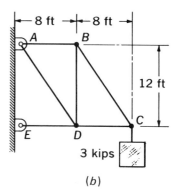
(b)

FIGURE 4.15   Problem 11

Framed Structures and Friction   93

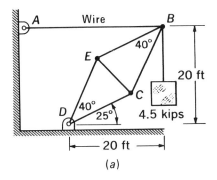

(a)

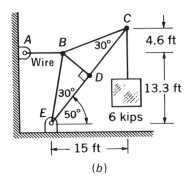

(b)

FIGURE 4.16   Problem 12

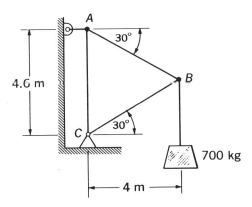

FIGURE 4.17   Problem 13

CHAPTER FOUR

15. Use the A-frame in Sample Problem 4.3, but change the loading. Place one 500 lb load on ABC, located horizontally 6 ft from A, and place another 500 lb load on CDE, located horizontally 6 ft from E.
16. Determine the forces applied to each member of the frame shown in Figure 4.18.
17. Repeat Problem 16, but use Figure 4.19.
18. Repeat Problem 16, but use Figure 4.20.
19. Repeat Problem 16, but use Figure 4.21.
20. Repeat Problem 16, but use Figure 4.22.

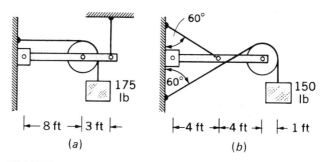

FIGURE 4.18 Problem 16

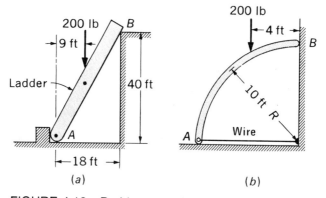

FIGURE 4.19 Problem 17

Framed Structures and Friction 95

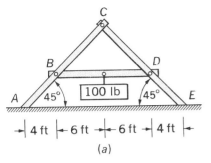
(a)

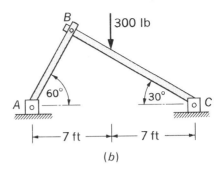
(b)

FIGURE 4.20 Problem 18

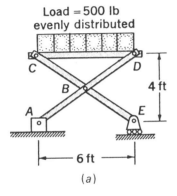
(a)

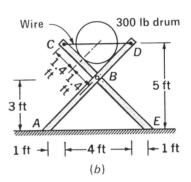
(b)

FIGURE 4.21 Problem 19

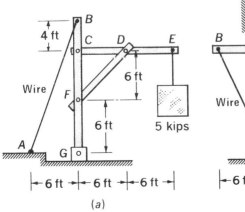
(a)

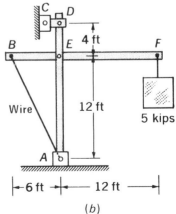
(b)

FIGURE 4.22 Problem 20

96 CHAPTER FOUR

21. Repeat Problem 16, but use Figure 4.23.
22. (SI Units) Refer to Figure 4.19(a). Change the feet to metres and the weight to 345 kg. Solve for the forces acting on member AB. Remember, force units must be in newtons.

Friction

23. A metal block weighing 500 lb is pushed across a level oak floor. What is the angle of friction? Determine the force F necessary just to start motion of the metal block, if F is applied (a) parallel to the floor, (b) downward at an angle of 30° to the floor, and (c) upward at an angle of 30° to the floor. Use the average value for the coefficient.
24. A 10 lb aluminum block is placed on a 30° incline.
    (a) What is the angle of friction?
    (b) Will the block slide down the incline due to its own weight?
    (c) Will the block slide down a 60° incline due to its own weight?
25. A 3000 lb car has its brakes locked. The car is on a level surface. With reference to the contact surface between the tire and the road:
    (a) What is the angle of friction?
    (b) What horizontal force is necessary just to start motion?
26. A 60 lb block is placed on a 40° incline. The angle of friction is 20°.
    (a) What force parallel to the incline is necessary to prevent downward sliding of the block?
    (b) What force parallel to the incline is necessary just to start motion up the incline?
    (c) To prevent sliding down the incline, at what angle to the horizontal should the force be placed in order to be a minimum value?
    Hint: In the force diagram, decide the direction of the restraining force to provide the shortest distance between the weight vector and the vector of the single force of resistance provided by the incline.
27. Refer to Figure 4.24. The angle of friction for all surfaces is 15°. Determine what the weights of the blocks marked W must be in order to prevent the 100 lb blocks from sliding down the inclines.
28. Using the information in Figure 4.25 and the facts that the angle of friction for all surfaces is 10° and the weight of block A is 200 lb, determine what the weight of block W must be in order to prevent motion.

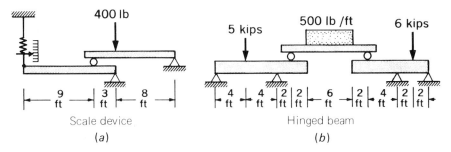

FIGURE 4.23 Problem 21

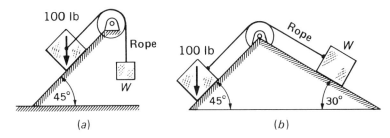

FIGURE 4.24 Problems 27 and 30

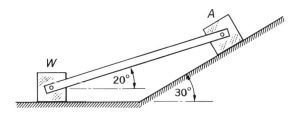

FIGURE 4.25 Problem 28

29. An eraser is being held against a blackboard with a horizontal force. If the eraser weighs 1 lb and the coefficient of friction is 0.7, what force is required?
30. (SI Units) Refer to Figure 4.24(a). What is the weight of the block on the incline (in kilograms) if the force in the rope is 50 N?

Wedges

31. Refer back to Figure 4.8. Block M weighs 100 lb, bar K is weightless, and the coefficient of friction is 0.4 between K and M. Consider the floor frictionless.

(a) Determine P necessary just to start sliding.
(b) Do part (a) above, but consider the floor to have a coefficient of friction of 0.4.
(c) Do part (b), but consider the weight of K to be 100 lb.

32. Refer back to Figure 4.8. Replace the bar K with a wedge having a 10° incline, so that block M moves up when the wedge moves to the right. Notice that there are ball bearings between the wall and block M, so that the wall offers no frictional force. The angle of friction for all other surfaces is 30°. The block and wedge each weigh 200 lb. Determine the force P just to start sliding motion.

33. Refer back to Figure 4.9. We are given the following information: The angle of friction for all surfaces is 30°. The wedge incline is 15°. The block and wedge each weigh 200 lb.
(a) Determine the force P to start sliding motion.
(b) What is the vertical force N supplied by the floor?

CASE STUDY

The roof framing of a house is quite often similar to an *A*-frame. You are a contractor making *A*-frames to the dimensions given in Sample Problem 4.3. One customer wants more head room in her attic and asks that the vertical height from A to C be changed to 10 ft (the span between A and E remains unchanged). Neglecting the weight of the frame, and using the loading specified in Problem 15, determine whether or not the pins at A, B, C, D, and E need to be strengthened because of increased forces.

COMPUTER PROGRAM

The following program will calculate two unknown forces needed for equilibrium of a truss joint. Any number of members can be connected to the joint as long as only two are unknown. All angles must be known and must be measured CCW from the plus *X*-axis. All tensile forces must be given a positive sign, and all compressive forces must be given a negative sign, regardless of direction.

The program sums the *X* and *Y* components of the known forces and places these in the equilibrium equations with the unknown components. Solution for the two unknown forces is by determinants.

```
10   REM   JOINT
20   PRINT "METHOD OF JOINTS FOR TRUSS BRIDGE.  PROGRAM WILL SOLVE FOR 2 U
     NKNOWN FORCES AT A PINNED JOINT.  USE '+' FOR TENSILE AND '-' FOR CO
     MP. FORCES.  MEASURE ANGLES   CCW   FROM THE X-AXIS."
25   PRINT
30   PRINT "TYPE IN THE NUMBER OF KNOWN FORCE MEMBERS."
40   INPUT N
60   PRINT
70   PRINT "TYPE IN THE AMOUNT OF FORCE ON EACH KNOWN MEMBER."
80   FOR I = 1 TO N
90   INPUT F(I)
100  NEXT I
105  PRINT
110  PRINT "TYPE IN EVERY MEMBER'S ANGLE TO THE X-AXIS--IN THE SAME SEQUE
     NCE AND INCLUDE UNKNOWN FORCE MEMBERS."
120  FOR I = 1 TO (N + 2)
130  INPUT A(I)
140  NEXT I
180  REM   --  CONVERT DEGREES TO RADIANS.--
190  LET Z = 3.1416 / 180
200  REM  -- SUM 'X' COMPONENTS OF KNOWN FORCES.--
210  FOR I = 1 TO N
220  LET X(I) = F(I) * ( COS (A(I) * Z))
230  NEXT I
235  LET C1 = 0
236  REM  -- C1 AND C2 ARE NEGATIVE BECAUSE THEY ARE MOVED ACROSS '=' SIG
     N IN EQUATION."
240  FOR I = 1 TO N
250  LET C1 = (C1 + X(I))
260  NEXT I
265  LET C1 =  - C1
270  REM  -- SUM 'Y' COMPONENTS.--
280  FOR I = 1 TO N
290  LET Y(I) = F(I) * ( SIN (A(I) * Z))
300  NEXT I
310  LET C2 = 0
320  FOR I = 1 TO N
330  LET C2 = (C2 + Y(I))
340  NEXT I
345  LET C2 =  - C2
350  LET A1 =  COS (A(N + 1) * Z)
360  LET A2 =  SIN (A(N + 1) * Z)
370  LET B1 =  COS (A(N + 2) * Z)
380  LET B2 =  SIN (A(N + 2) * Z)
390  LET DEN = (A1 * B2) - (A2 * B1)
400  LET XX = ((C1 * B2) - (C2 * B1)) / DEN
410  LET YY = ((A1 * C2) - (A2 * C1)) / DEN
420  PRINT
430  PRINT
440  PRINT "AT ANGLE ";A(N + 1);" DEG.  THE FORCE = "; INT (XX * 10 + .51
     ) / 10;" LB."
445  PRINT
450  PRINT "AT ANGLE ";A(N + 2);" DEG.  THE FORCE = "; INT (YY * 10 + .51
     ) / 10;" LB."
460  END
```

5

Centroids and Moments of Inertia of Areas

OBJECTIVES

This discussion of centroids and moments of inertia is intended as a review. The concepts are demonstrated and explained, and procedures for solving problems are covered. Upon completion of this chapter, you will be able to locate the centroid of composite areas formed by various combinations of circles, rectangles, and triangles. You should be able to apply the proper moment of inertia and radius of gyration formulas for circles and rectangles. You will also be able to add and subtract moments of inertia taken about the same axis.

INTRODUCTION

This chapter covers an important area of statics. The information aids in the understanding of equilibrium conditions. It is also helpful for a thorough understanding of strength of materials, especially the strength of beams and columns.

CENTROID—FIRST MOMENT OF AREA

The gravitational force of the earth acting on the object in Figure 5.1(*a*) is actually exerting a force on each atom in the object. These are all equal parallel forces and (as mentioned in Chapter 3) may be replaced by an equivalent single force, as in Figure 5.1(*b*). This single force acts at the *center of gravity* (cg) of the object. The cg is the point where the

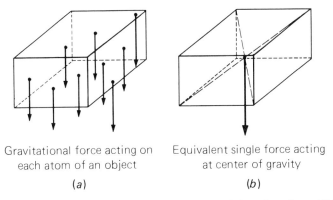

Gravitational force acting on each atom of an object
(a)

Equivalent single force acting at center of gravity
(b)

FIGURE 5.1 Location of the Center of Gravity of an Object

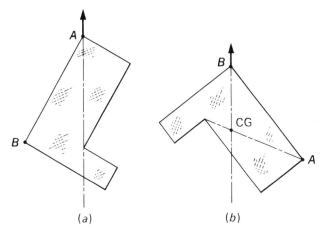

(a)   (b)

FIGURE 5.2 Experimental Determination of the Center of Gravity for a Thin Irregular Object

moments of the weights of the atoms on one side balance the moments of the atoms on the other side:

$$\Sigma M = 0$$

If an object has a regular geometric shape, such as cube, parallelepiped, or sphere, the cg may be determined easily because it is at the geometric center of the shape (assuming even distribution of the material). If the shape is irregular, the problem is more difficult. An experimental method that can be used with thin objects is illustrated in Figure 5.2. Hang the object from point A, and inscribe a vertical line directly under A. Now hang the object from some other point (B in the

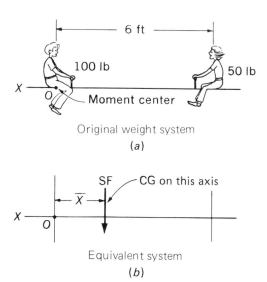

FIGURE 5.3 Determination of the Vertical Axis of the Center of Gravity by Locating the Position of the Equivalent Single Force

illustration), and inscribe another vertical line under B. The intersection of the two lines indicates the cg. If the object in Figure 5.2 were infinitely thin (an area only), it could have no weight; technically there could be no cg. The point where the cg *normally* would be must have a name; in the past, cg and center of area have been used. The term in common use today is *centroid*.

The centroids of regular geometric areas, such as rectangles, triangles, and circles, are easily located at their geometric centers. If the area has an irregular shape, the method described in the following paragraphs is useful for locating the centroid.

Prior to working with areas, we will review the method for finding the cg of two solids. The cg is the balancing point, because the moments of all forces on one side must balance the moments of all forces on the other side. Figure 5.3 shows two children on a seesaw to illustrate the method of finding the cg. The seesaw itself is considered weightless to simplify the problem. The question is, "Where does the fulcrum go so the seesaw will balance?" The fulcrum must go directly under the cg of the combined weights of the two children. To locate the cg, the two children's weights are replaced by an equivalent single weight (single force) acting through the cg. If this single force is to be equivalent, it must have the magnitude and the same moment (about the same moment center) as the original system. Expressed algebraically:

$$SF \cdot \overline{X} = \Sigma w \cdot x$$
$$\overline{X} = \frac{\Sigma w \cdot x}{SF}$$

where
SF = equivalent single force
$\overline{X}$ = distance from the chosen moment center to SF
w = weight of each object
x = distance from the chosen moment center to the center of each weight

The calculations are as follows:

$$SF = \Sigma F = -100 \text{ lb} - 50 \text{ lb} = -150 \text{ lb}$$
$$SF \cdot \overline{X} = \Sigma w \cdot x$$
$$150 \text{ lb} \cdot \overline{X} = 50 \text{ lb} \cdot 6 \text{ ft}$$
$$\overline{X} = 2 \text{ ft}$$

We use the same procedure for determining the centroidal axes of composite areas. The only difference is that areas (usually in square inches) are used instead of weights or forces. Look at Figure 5.4. The indirect question is, "Where should the centroid of an equivalent single area be placed?" Choosing reference axis Y through the centroid of area 1 and moment center O on axis Y, we sum the moments of the areas as if they were forces. Expressed algebraically:

$$A \cdot \overline{X} = \Sigma a \cdot x$$
$$\overline{X} = \frac{\Sigma a \cdot x}{A}$$
$$A \cdot \overline{Y} = \Sigma a \cdot y$$
$$\overline{Y} = \frac{\Sigma a \cdot y}{A}$$

where
A = total composite area
$\overline{X}, \overline{Y}$ = respective distances from the chosen moment center to the centroid of the composite area
a = designation for each individual area in the composite area
x, y = respective distances from the chosen moment center to the centroids of the individual areas

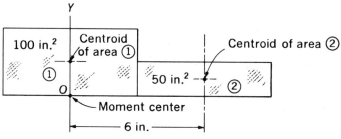

Combined area showing centroid for each area
(a)

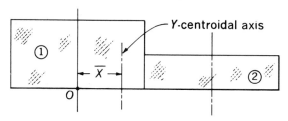

Illustration of Y-centroidal axis of combined area
(b)

FIGURE 5.4 Determination of the Centroidal Y Axis of the Combined Area

We will go through the steps for finding the x distance:

$A = \Sigma a = 100 \text{ in.}^2 + 50 \text{ in.}^2 = 150 \text{ in.}^2$
$A \cdot \overline{X} = \Sigma a \cdot x$
$150 \text{ in.}^2 \cdot \overline{X} = 50 \text{ in.}^2 \cdot 6 \text{ in.}$
$\overline{X} = 2 \text{ in.}$

Procedure for Solving Centroid Problems

Step 1 Draw a sketch of the area, and choose a convenient location for the moment center.

Step 2 Divide the area into standard geometric shapes, and locate the centroid for each individual area. If necessary, refer to the table of properties of standard geometric areas in the Appendix.

Step 3 Substitute values in the equations above, and solve for $\overline{X}$ and $\overline{Y}$.

Sample Problem 5.1 indicates the steps used to locate the centroid of a composite area.

SAMPLE PROBLEM 5.1
Finding the Centroid of a Composite Area

Problem

Determine the position of the centroid of the composite area shown in the following sketch. Dimensions are in inches.

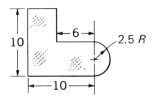

Solution

Step 1 The moment center O is chosen for convenience at the bottom left-hand corner of the area.

Step 2 The composite area is divided into standard geometric shapes, and the x and y distances to the centroid of each shape are established. From the Appendix, the centroid of a semicircle is located $0.424 \times R = 0.424 \times 2.5$ in. $= 1.06$ in. from the base.

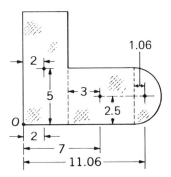

Step 3 Solve for $\overline{X}$ and $\overline{Y}$, using Equations 1 and 2 given earlier in the chapter. *Note:* The area of the semicircle is 9.8 in.

$$\overline{X} = \frac{\Sigma a \cdot x}{A} = \frac{(40 \cdot 2) + (30 \cdot 7) + (9.8 \cdot 11.06)}{40 + 30 + 9.8} = 5 \text{ in.}$$

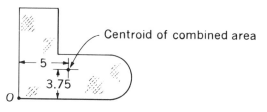

$$\overline{Y} = \frac{\Sigma a \cdot y}{A} = \frac{(40 \cdot 5) + (30 \cdot 2.5) + (9.8 \cdot 2.5)}{79.8} = 3.75 \text{ in.}$$

It is interesting to note that the centroid of an area need not be located on the area (look ahead to Figures 5.7 and 5.8). Centroids are called the *first moment of area* because they are determined by multiplying the moment distance (raised to the first power) by the area. This is in contrast to the method of determining the moment of inertia described next.

MOMENT OF INERTIA OF AN AREA—SECOND MOMENT OF AREA

The moment of inertia of an area is an abstract value used to help solve various engineering problems. The term *abstract* means that no part of the area can be identified as the moment of inertia. This is different from the centroid, which is a concrete value and can be located with respect to the area. The term *moment of inertia* is used because its purpose is analogous to that of the dynamic moment of inertia. In concept, the moment of inertia of an area is equal to the area multiplied by the distance from the centroid to the chosen moment center, squared:

Concept formula: $I = A \cdot d^2$

where

I = moment of inertia in inches to the fourth power
A = area
d = distance from the chosen moment center to the centroid

The distance d also may equal $\overline{X}$ and $\overline{Y}$. *Note:* This formula represents a concept only; it is *not* a formula to be used.

An example of the moment of inertia is shown in Figure 5.5. The proportions of area to distance from the moment center were cho-

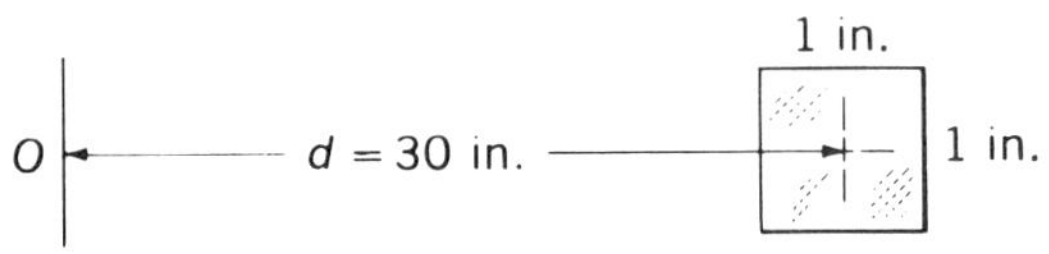

```
                           1 in.
                       ┌─────┐
O |——— d = 30 in. ———  │  ·  │ 1 in.
                       └─────┘
```

$I = A(d)^2 = 1(30)^2 = 900$ in.^{4}

FIGURE 5.5 Moment of Inertia—Concept Only

sen so that the answer of 900 in.^{4} is very close to the true answer. But for the problems in this text, the concept can only be applied with the aid of calculus. Calculus has already been used to give us the algebraic formulas for specific areas; these formulas are listed in the Appendix in the table of properties of geometric shapes. Also, for convenience, we speak of the moment of inertia being taken about some specified moment *axis* rather than moment *center*. Note that the moment of inertia units are inches^4.

Polar Moment of Inertia

Figure 5.6 indicates that the centroid of an area can be located on three main axes: X, Y, and Z. The Z-axis is perpendicular to the plane of the area. A moment of inertia taken about the Z-axis is called the *polar moment of inertia*. Its symbol is J. This will be needed for solving circular shaft problems in Chapter 12.

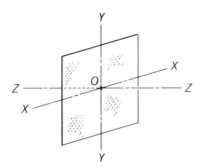

FIGURE 5.6 Centroidal Axes X, Y, and Z Pass through the Centroid

Radius of Gyration

The radius of gyration is another abstract value, used primarily in the solution of column problems (see Chapter 19). It is equal to the square root of the moment of inertia, taken about the X or Y centroidal axis, divided by the area. Stated as a formula:

$$r = \sqrt{\frac{I}{A}}$$

where

r = radius of gyration in inches

The radius of gyration can be thought of as the d value in the *concept formula* of $I = A \cdot d^2$. If I is about the X centroidal axis, then r is considered to be taken about the X centroidal axis.

Procedure for Finding Moment of Inertia or Radius of Gyration of an Area

Step 1 Locate the proper formula in the table of properties of standard geometric areas in the Appendix.

Step 2 Substitute the given values.

Sample Problem 5.2 shows how straightforward this procedure is.

SAMPLE PROBLEM 5.2
Moment of Inertia and Radius of Gyration for a Rectangular Area

Problem

Obtain the true dimensions of the cross section of a 2 × 4 timber and find (a) the moment of inertia about the X centroidal axis, I_x, (b) the I_y value, and (c) the radius of gyration about the X centroidal axis, r_x.

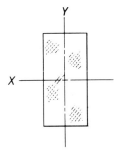

Solution

Step 1 From the Appendix table of lumber sizes, a 2 × 4 is 1½ × 3½. From the Appendix table of geometric shapes,

$$I = \frac{bh^3}{12}.$$

Step 2 By substitution,

(a) $I_x = \dfrac{1.5(3.5)^3}{12} = 5.35$ in.^{4}

(b) $I_y = \dfrac{3.5(1.5)^3}{12} = 0.985$ in.^{4}

(c) $r_x = \sqrt{\dfrac{5.35}{1.5(3.5)}} = \sqrt{1.02} = 1.01$ in.

SUMMARY

Locating the centroid of a composite area is analogous to finding the center of gravity of a number of weights and involves replacing the given areas or force system with an equivalent single area or force.

The concept of the moment of inertia of an area is related to the concept of locating the centroid, in that the moment arm value is squared. For this reason, some authorities refer to the moment of inertia as the *second moment of area*. The moment of inertia is an abstract concept ("inches raised to the fourth power" has no physical meaning) and is an aid to the solution of problems concerning torque on shafts and bending moments in beams and columns in this text.

PROBLEMS

Centroids

Note: The X- and Y-axes refer to the centroidal axes.

1. Determine the position of the centroid for the composite area in Figure 5.7.
2. Repeat Problem 1, but use Figure 5.8.
3. Repeat Problem 1, but use Figure 5.9.
4. Repeat Problem 1, but use Figure 5.10.
5. Repeat Problem 1, but use Figure 5.11.
6. Figure 5.12 shows a riveted plate. Locate the centroid of the rivet areas.
7. Figure 5.13 also shows a riveted plate. Locate the centroid of the rivet areas for this problem.
8. (Extra Credit) How high must H be so that the centroid of the combined area lies on the X-axis shown in Figure 5.14?
9. (Extra Credit) Given the information in Figure 5.15, replace the distributed load with a single force through the cg. Use Simpson's Rule, graphical calculus, or information from the Appendix regarding properties of geometric shapes (parabolas).

Moment of Inertia

Note: Refer to the table of properties of geometric shapes in the Appendix. Axes are centroidal axes.

10. Determine the I_x and I_y of a rectangle 9 in. high and 6 in. wide.
11. Determine the I_x of an isosceles triangle with a 3 in. base and a 9 in. height.

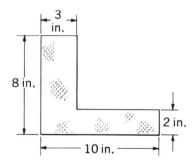

FIGURE 5.7   Problem 1

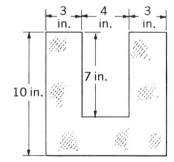

FIGURE 5.8   Problem 2

Centroids and Moments of Inertia of Areas 111

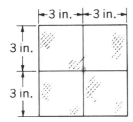

FIGURE 5.9   Problem 3

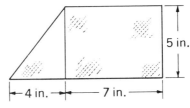

FIGURE 5.10   Problem 4

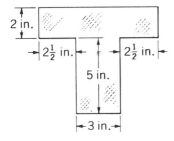

FIGURE 5.11   Problem 5

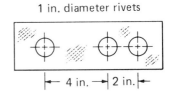

FIGURE 5.12   Problem 6

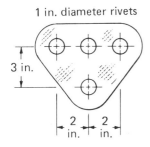

FIGURE 5.13   Problem 7

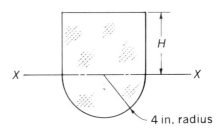

FIGURE 5.14   Problem 8

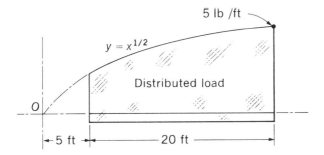

FIGURE 5.15   Problem 9

12. Determine the I_x and the I_y of an annular area with an inside diameter of 3 in. and an outside diameter of 4 in.

Radius of Gyration

13. Determine the r_x of the area described in Problem 10.
14. Determine the r_x of the area described in Problem 11.
15. Determine the r_x of the area described in Problem 12.

6

Stress, Strain, and Deformation

OBJECTIVES

This chapter defines some basic terms in strength of materials and introduces you to the formulas for stress and strain. You will learn how to apply these formulas to various structures and machine parts with forces distributed evenly or concentrated along a central axis.

INTRODUCTION

The term *strength of materials* refers to the *mechanical* changes (as opposed to chemical, electrical, or other changes) occurring in a material when an external force is applied. You must know that the design of a machine part or structure component will be strong enough to perform its function satisfactorily, but not to the point where the machine or structure lacks economy or exceeds space limitations. Also, there are deflection and twisting considerations. For example, the designer of torsion bar suspensions for cars must know how much twist is required for a particular applied force. The designer of a turbine generator set bed must know the deflection limit, so that the shafts of the turbine and generator remain straight and do not join at an excessive angle.

When a force is applied to a material that remains in equilibrium, the material must resist the force. If the force is gradually increased, a point may be reached where the material can no longer resist, and so it ruptures. The behavior of a material under load is our topic for the rest of this text. Studies of machine elements or structures will be restricted to those in equilibrium. The forces will be considered constant or applied gradually.

You should consult other texts if you wish to investigate the behavior of materials that are accelerated or subjected to sharply varying loads.

TENSILE, COMPRESSIVE, AND SHEAR STRESSES

Figure 6.1 illustrates a steel bar with a 10 000 lb force applied to end A. From our study of statics in the previous chapters, we know that if a force is applied to a body that remains in equilibrium, then an equal force must oppose the original force. This resisting force is applied at point B in Figure 6.1. An imaginary cut is made at section C–C, shown in the figure, and an FBD of the left half of the bar is drawn. Because the whole steel bar is in equilibrium, every part (such as the left half) must be in equilibrium, and the internal resisting force equals 10 000 lb, as shown in Figure 6.1(b).

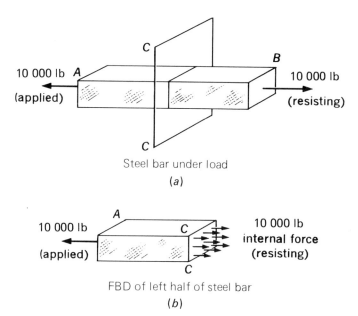

FIGURE 6.1 Steel Bar under Load and in Equilibrium

An internal *tensile* force occurs in a material when two externally applied forces are equal, opposite, and collinear and tend to *pull* the material. An internal *compressive* force occurs in a material when two externally applied forces are equal, opposite, and collinear and tend to *push* or *squeeze* the material. An internal *shear* force develops in a material when two externally applied forces are equal, opposite, and *closely parallel* to each other. For example, place a deck of cards between the palms of your hands and slide your hands apart. The force that causes one card to slide over another is a shear force. Figure 6.2 illustrates these internal forces. Internal forces resulting from more complicated force systems are discussed in later chapters.

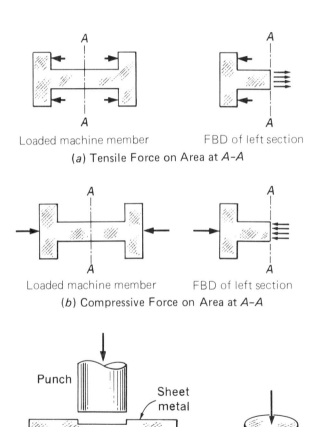

FIGURE 6.2 Tensile, Compressive, and Shear Forces

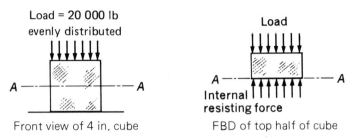

FIGURE 6.3 Four Inch Cube with Compressive Load

Figure 6.3 shows a cube, 4 in. on a side, supporting a compressive load of 20 000 lb. This load is distributed evenly across the surface of the block. Actually the whole block is supporting the load, but for our engineering calculations, only an area is considered to be supporting the load. If we were to investigate a surface on the side of the block, parallel to the applied force, we could see that no force is applied. However, a force is applied to the top surface of the cube and every surface (or area) parallel to the top surface. The most important area (called the critical area) resisting simple compressive or tensile forces is *perpendicular* to those forces. The critical area resisting simple shear forces is *parallel* to those forces.

Internal forces are considered to be evenly distributed across the areas they are acting on. Engineers and technicians must know the force acting on each square inch of area; the term used to describe this force is *stress*. The units of stress are lb/in.² or psi. The symbol for tensile or compressive stress is the Greek letter *sigma*, σ, and for shear stress the symbol is the Greek letter *tau*, τ. The formula is:

$$\sigma \text{ (or } \tau\text{)} = \frac{F}{A}$$

where

σ or τ = stress in psi (pascals)
F = force in lb (newtons)
A = area in square inches (square metres)

When using this formula, it is assumed that the stress in a body is uniformly distributed across the area resisting the force. There are situations when this is not true; they are discussed later in the text. Also, it is assumed that the material is homogeneous and that there are no air bubbles or dense spots to distort the stress pattern. A typical situation is presented in Sample Problem 6.1.

Stress, Strain, and Deformation   117

SAMPLE PROBLEM 6.1
Stresses on Critical Areas

Problem

In an experiment to determine the strength of a glue, the three blocks (A, B, and C) are glued together as shown in the following sketch. When the compressive load of 11 000 lb is applied, determine the stresses on the critical areas.

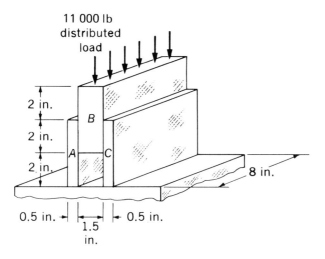

Solution

Step 1 Refer to sketches (a) and (b). The shear stress on the glued area of block B is calculated as follows:

$$\tau = \frac{F}{A} = \frac{5500 \text{ lb}}{2 \text{ in.} \times 8 \text{ in.}} = 344 \text{ psi}$$

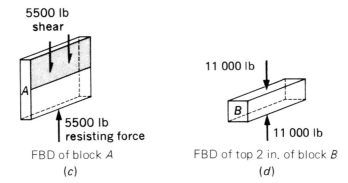

FBD of block A
(c)

FBD of top 2 in. of block B
(d)

Note: Since all the shear areas are the same, all the shear stresses have the same value. Consequently, the shear stress on the glued area of block A, as shown in sketch (c), is also 344 psi.

Step 2 Refer to sketch (d). Compressive stress on an area 2 in. below the top surface of block B is:

$$\sigma_c = \frac{F}{A} = \frac{11\,000}{1.5 \text{ in.} \times 8 \text{ in.}} = 916 \text{ psi}$$

Step 3 The FBD for the following solution has been left for you to do. Compressive stress on an area 2 in. above the bottom surface of block A is:

$$\sigma_c = \frac{F}{A} = \frac{5500}{0.5 \text{ in.} \times 8 \text{ in.}} = 1375 \text{ psi}$$

This answer also applies to block C, since it has the same forces applied and the same dimensions.

DEFORMATION

Humans have been building large structures and using metals for thousands of years but have understood the relationship of applied loads and deformation of supporting structures for only a few hundred years. The English scientist Robert Hooke (1635–1703) first demonstrated the relationship.

When an external force acts on a machine part, it deforms the part. The material tends to resist the deformation and set up stresses. The deformation of a machine part under load is quite important and

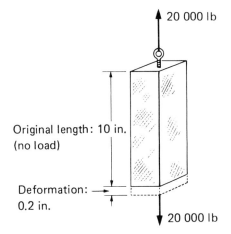

FIGURE 6.4 Uniform Bar under Load

must be considered frequently in designs. In Figure 6.4, a tensile load of 20 000 lb is applied to a uniform 10 in. bar, and the bar is observed to increase 0.2 in. in length. This deformation is represented by the Greek letter *delta*, δ. The deformation for each inch of length is called *strain*. For the bar, the strain is equal to:

$$\frac{0.2 \text{ in.}}{10 \text{ in.}} = 0.02 \text{ in./in.}$$

The symbol for strain is the Greek letter *epsilon*, ε. The letter L is used to designate the original length of the machine member. Using these symbols, the general form of the equation becomes a second important formula in strength of materials:

$$\epsilon = \frac{\delta}{L}$$

where
  δ = deformation in inches (metres)
  L = original length in inches (metres)
  ε = strain, the deformation for each inch (metre) of length

The units of strain are inch/inch or metre/metre. Although, mathematically, strain is dimensionless, the units are often included for clarity.

For most engineering materials, the amount of deformation a machine part undergoes is directly proportional to the force applied.

Each material has a limit to this proportional deformation, as will be discussed in detail in Chapter 7. As an example of what is meant by proportional deformation (or strain), suppose that the tensile force on the bar in Figure 6.4 is increased to 40 000 lb. The new length of the bar would be 10.4 in. When the load doubles, the deformation (and strain) doubles. This fact was first noted by Robert Hooke in about 1676, and is frequently referred to as Hooke's Law. Expressed mathematically, Hooke's Law becomes:

$$\delta \propto F$$

which is read: *deformation varies directly as the force.* If a specimen of unit size (1 sq. in. in area and 1 in. in length) is used, the above proportionality can be written:

$$\epsilon \propto \sigma$$

which is read: *strain varies directly as the stress.*

The next advance was made by Thomas Young (1773–1829), an English physicist and physician. In 1802, Young made further observations concerning Hooke's Law. Let us demonstrate Young's observations with a specific example.

A specimen is placed under a stress of 30 000 psi, and the strain is measured as 0.001 in./in. With this information the proportionality above becomes:

$$0.001 \text{ in./in.} \propto 30\,000 \text{ psi}$$

This proportionality means that if the stress were doubled to 60 000 psi, the strain would double to 0.002 in./in. Young decided to make this expression an equality by multiplying the strain by a constant.

$$0.001 \text{ in./in.} \times (30\,000\,000 \text{ psi}) = 30\,000 \text{ psi}$$

Young noticed that this constant was indeed the same value under various conditions of length, cross-sectional area, and load for the same material. Each material has its own constant. Engineering societies have agreed to call this constant the *modulus of elasticity*, and its symbol is E. Young's formula relates stress to strain. This relationship is for tensile and compressive stresses. The *modulus of elasticity in shear* will be discussed in a later chapter. The general form of Young's formula is as follows:

$$\epsilon \times E = \sigma$$

or more often:

$$E = \frac{\sigma}{\epsilon}$$

where

σ = stress in psi (pascals)
ϵ = strain (dimensionless)
E = modulus of elasticity in psi (pascals)

Note from the formula that since E is constant for a given material, any change in stress results in a proportional change in strain. For example, cut the stress in half, and the strain value is reduced to half.

The modulus of elasticity for various engineering materials is given in tabular form in the Appendix. You should be alert that all materials may not necessarily follow Hooke's Law; however, the engineering materials we will deal with follow his law closely enough for practical purposes. Sample Problems 6.2 and 6.3 (metric) deal with strain.

Procedure for Solving Stress/Strain Problems

Step 1 Determine whether a force, area, stress, or strain is to be the key to the answer.

Step 2 Pay particular attention to your selection of areas to be investigated. Frequently, determining the proper area is the key to the solution.

Step 3 Draw an FBD of the complete assembly; then draw FBDs of the sections, separating the sections at the areas to be investigated.

Step 4 Using equations in the text thus far, solve the problem.

Step 5 Again, remember the checking procedure spelled out in Chapter 1.
   (a) Round off your figures for easy mental approximation of the answer.
   (b) Run through the equation(s) using dimensional units to make sure you do not obtain, for example, psi when you should have lb.
   (c) Check to see that your answer appears reasonable. For instance, if you know that structural steel cannot take more than 70 000 psi, then any answer above this figure points to an error in design or solution.

SAMPLE PROBLEM 6.2
Deformation and Strain

Problem

A steel tension rod member in a truss is 60 ft long and has a cross-sectional area of 1.5 in.^{2}. (a) What is the strain when a 40 000 lb load is applied? (b) What is the elongation of the member with that load?

Solution (a)

E for steel is 30 000 000 psi (30×10^6 psi). Using Young's formula,

$$E = \frac{\sigma}{\epsilon}$$

$$\epsilon = \frac{\sigma}{E} = \frac{F/A}{E} = \frac{F}{A \cdot E} = \frac{40\ 000}{1.5(30 \times 10^6)} = 0.888(10^{-3}) = 0.000\ 89$$

Solution (b)

$$\delta = \epsilon \cdot L = 0.89(10^{-3}) \times 60 \text{ ft} \times 12 \text{ in./ft} = 0.639 \text{ in.}$$

METRIC SAMPLE PROBLEM 6.3
Stress and Deformation

Problem

A 45 mm diameter steel rod 10 m long supports a balcony. The tensile force on the rod is 300 kN. What is the stress and deformation? *Note:* Stress is in pascals (1 pascal = 1 newton/metre^2). Therefore, dimensions must be in metres:

45 mm = 0.045 m
300 kN = 300(10)^3 N

Solution

$$\sigma_t = \frac{F}{A} = \frac{300(10)^3}{\pi(0.045)^2/4} = 189(10)^6 \text{ Pa} = 189 \text{ MPa}$$

$$\delta = \frac{\sigma \cdot L}{E} = \frac{189(10)^6 \cdot 10}{200(10)^9} = 0.0095 \text{ m} = 9.5 \text{ mm}$$

SUMMARY

This chapter deals with forces applied along one axis only. That is, if an applied force is directed vertically downward, the reaction is vertically upward, and no force will be applied horizontally or at other angles to the part. Under these conditions, the maximum tensile or compressive stresses will appear on an area of the part that is perpendicular to the line-of-action of the force. Maximum shear stresses appear on areas parallel to the line-of-action of the force.

All materials deform when a force, or load, is applied. The materials we will be discussing deform in direct proportion to the load. Strain is the deformation for each inch or metre of length. Strain is dimensionless, but the units are often included for clarity. Calculations in stress and strain problems are based on the following formulas.

1. Formula for stress:

$$\sigma \text{ (or } \tau) = \frac{F}{A}$$

2. Formula for strain:

$$\epsilon = \frac{\delta}{L}$$

3. Formula relating tensile or compressive stress to strain:

$$E = \frac{\sigma}{\epsilon}$$

PROBLEMS

1. A steel block measuring 2 in. wide, 4 in. long, and 5 in. high supports a 175 000 lb distributed load (applied vertically downward).
   (a) What is the stress on a horizontal cross-sectional area?
   (b) If the area of the cross section were doubled, what would be the stress?
   (c) Is the stress on the horizontal area tension, compression, or shear?

(d) If the material of the block can take only a stress of 10 000 psi, how large an area is required?
2. A hollow round support post has a 3 in. outside diameter. The stress must not exceed 15 000 psi. If the distributed load is 50 000 lb, what must the inside diameter be?
3. A 1/2 in. diameter manila rope can support a safe load of 530 lb. If we assume that the fibres completely fill the 1/2 in. diameter circle, what stress will be on the rope's cross section?
4. A section of a shaft and thrust collar are shown in Figure 6.5. The collar is shrunk-fit to the shaft and resists a 700 lb force applied to the shaft.
    (a) What is the tensile stress on a cross section of the shaft?
    (b) What is the shear stress on the area of contact between the collar and shaft?
5. Figure 6.2(c) shows a 1 in. diameter punch and a 1/4 in. thick plate. The maximum stress to be placed on the punch is 70 000 psi compressive. The stress that causes shear failure in the plate is 45 000 psi.
    (a) What force must be applied to the punch to perform the punching operation?
    (b) Is the stress on the punch below the 70 000 psi limit?
    (c) How thick can the plate be if the 1 in. punch is stressed to its 70 000 psi limit?
    (d) If the plate thickness is kept at 1/4 in. but the punch diameter is varied, is there a limit to the punch diameter? If so, is the limiting diameter greater or less than 1 in.?
6. Figure 6.6 shows the cross section of a type of adjustable jack post to support sagging floors that can be obtained from hardware stores. The jack post has a recommended maximum load of 15 000 lb.
    (a) Determine the compressive stresses on sections *AA*, *BB*, and *CC*.
    (b) Determine the shear stress on the 7/8 in. diameter pin. (Note that for complete failure of the pin, two shear areas are involved.)
7. The steel box beam in Figure 6.7 supports an evenly distributed load of 150 000 lb. It is supported by a steel pin at one end; the other end is resting on a steel block mounted on a wood support. Steel may be loaded in compression to 17 000 psi and in shear to 10 000 psi. Wood may be loaded to a compressive stress of 500 psi.
    (a) Determine the size of the steel pin required for a shear load.

Stress, Strain, and Deformation 125

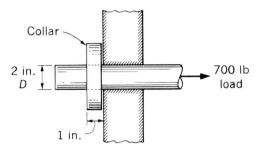

FIGURE 6.5 Problem 4

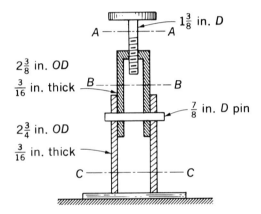

FIGURE 6.6 Problem 6

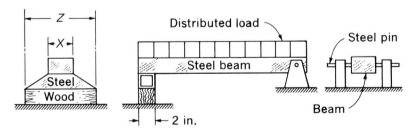

FIGURE 6.7 Problem 7

(b) Determine the distance x. The beam is resting on a 2 in. wide support, but the x dimension of the block has to be solved.

(c) Determine the distance z. The wood support is 2 in. wide, but the z dimension must be calculated.

8. A test specimen has a 2 in. gage length, marked as shown in Figure 6.8. When a 5000 lb tensile force is applied, the distance between gage marks is 2.030 in.

(a) What is the distance between marks with a 15 000 lb load? A 23 000 lb load?

(b) What strain results from each of the three applied loads?

9. In an experiment, a 20 ft piano wire elongated under load to 20.25 ft. What was the strain?

10. In Problem 1(a) above, assume the block is steel and is exactly 5 in. high with no load.

(a) What is the strain under the specified load?

(b) What is the height of the block under the specified load?

11. Brass has a modulus of elasticity of 16 000 000 psi. If a load elongates a 1 in. specimen 0.002 in.:

(a) What is the stress?

(b) If the cross-sectional area of the specimen is 0.45 in.^{2}, what load is applied?

12. A surveyor's 100 ft steel tape is 0.30 in. wide and 0.02 in. thick. If the tape measures exactly 100 ft when supported over its length with a 10 lb pull, what distance will 100 ft on the tape measure be when a 30 lb pull is applied? What is the tensile stress with a 30 lb pull?

13. Figure 6.9 illustrates a bolt and nut assembly holding two plates together. The shank diameter and thread minor diameter of the bolt are each 3/4 in. The nut has been run up snugly, so there is neither slack nor stress in the bolt. If the nut is rotated 1/4 turn, what stress develops in the bolt? Assume the only deformation is that of the bolt lengthening. Solve for a 1020 CD steel bolt, a brass bolt, and an aluminum bolt. Refer to the Appendix for the modulus of elasticity of each material.

14. (SI Units) A square aluminum rod 15 mm on a side and 2 m long undergoes a deformation of 5.2 mm under a tensile load. What is the load in newtons?

15. (SI Units) A steel tension rod 20 m long has a cross-sectional area of 50 mm^2. A 200 kN tensile load is applied.

(a) Give the strain.

(b) Give the deformation in mm.

Stress, Strain, and Deformation 127

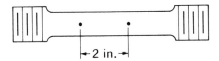

FIGURE 6.8   Problem 8

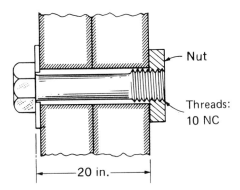

FIGURE 6.9   Problem 13

CASE STUDY/COMPUTER PROGRAM

Your company is designing the truss bridge shown in Figure 6.10. There are 16 bottom chord members, each 20 ft long (total length of truss = 320 ft). For reasons of economy, appearance, and design, your department head wants to know what the total change in length will be when

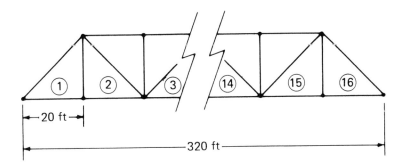

FIGURE 6.10   Truss Bridge for Computer Problem

the truss is loaded. Each lower chord member has a tensile load that is specified below. Choose an A-36 steel wide-flange beam (W-shape) that will safely take the maximum specified load. Do not exceed 22 000 psi tensile stress. Now, provide answers two ways:

1. Make all of the lower chord members the same size. Use the computer program to calculate the total deformation for the 320 ft. This, of course, means that most of the members are stressed well below the allowable stress.
2. Choose each lower chord member so that it is stressed as close to the allowable stress as possible, and calculate the total deformation for the 320 ft.

Member Number	Load in Pounds
1, 16	300 000
2, 15	400 000
3, 14	500 000
4, 13	600 000
5, 12	700 000
6, 11	800 000
7, 10	900 000
8, 9	1 000 000

```
10   REM  DEFORMATION
15   HOME
20   PRINT "THIS PROGRAM WILL CALCULATE THE DEFORMATION OF EACH PART OF A
     STRUCTURE,"
21   PRINT "LOADED IN EITHER ALL TENSION OR ALL COMPRESSION, AND TOTAL THE
     RESULTS."
25   LET TD = 0
30   PRINT "TYPE IN THE NUMBER OF MEMBERS TO BE EVALUATED."
40   INPUT N
50   FOR X = 1 TO N
60   PRINT "TYPE IN LOAD (LB), LENGTH (FT), AND CROSS SECTIONAL AREA (SQ.I
     N.)."
70   INPUT F,L,A
80   LET D =   INT ((F * L * 12 / (30E6 * A)) * 1000 + .51) / 1000
90   LET TD = TD + D
100  PRINT    TAB( 15)"DEFORMATION = ";D;" IN."
110  PRINT    TAB( 15)"TOTAL DEF. = ";TD;" IN."
120  PRINT
130  NEXT X
140  END
```

7

Mechanical Properties of Materials

OBJECTIVES

This chapter defines and explains a number of mechanical properties important in the field of strength of materials. As you learn the mechanical strengths and weaknesses of various materials, especially steel, you will be able to recognize the relative importance of various properties for specific design requirements. You will be able to take appropriate data, plot stress–strain curves, and interpret the information for design purposes.

INTRODUCTION

The famous Damascus steel swords of the Middle Ages were reputed to be able to cut a floating feather or cut down an armored opponent. Hard enough to hold a sharp edge, tough enough to take hard use, these swords had desirable mechanical properties only the rich could afford. Today, we rarely give a second thought to equivalent steel products for everyday use in band saws, hack saws, and machine tool bits. Engineering materials, particularly steel, can often be "tailored" to the type of job they will be performing.

Materials have numerous properties, such as chemical, thermal, electrical, and magnetic, but the discussion in this text will be on the mechanical properties or characteristics. Investigating mechanical properties is important because of new materials and new environmental conditions in space activity, high temperatures in power generation equipment, low temperatures in arctic areas, and high pressures under the sea. The materials discussed in this chapter are those generally used to support loads as members of machines or structures.

STATIC TENSILE TEST

Figure 7.1 shows a universal testing machine. This machine can apply tensile, compressive, shear, or bending loads to test specimens. The *static tensile test* is one of the most fundamental tests for determining

FIGURE 7.1 Reliability of a Railroad Coupling Being Analyzed with a 1 200 000 lb Universal Testing Machine

a number of mechanical properties. A standard ASTM (American Society for Testing and Materials) test specimen is made according to the information in Figure 7.2 and placed in the machine. The load on the specimen is increased slowly because this is considered to be a static condition (as compared to an impact load). Loads are read from the testing machine's dial, and deformations are read from an extensometer fastened to the specimen, as illustrated in Figure 7.3.

Data for a typical specimen of steel are shown in Table 7.1. Notice that the load readings and elongation readings are converted to stress and strain respectively. From this data, a stress–strain curve is plotted with stress marked off on the ordinate and strain on the abscissa (see Figures 7.4 and 7.5). The initial portions of the curves are usually straight lines; this is the range (stress proportional to strain) in which Hooke's Law applies.

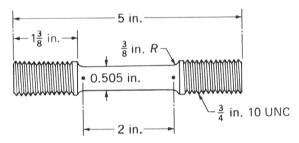

FIGURE 7.2   Tensile Test Specimen

FIGURE 7.3   Extensometer Mounted on Tensile Test Specimen

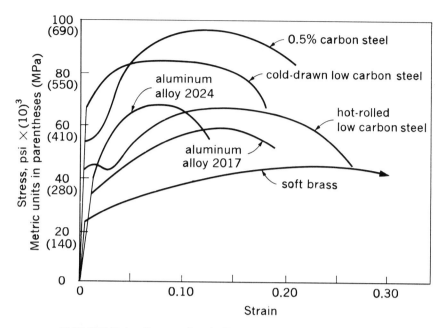

FIGURE 7.4 Stress–Strain Diagrams

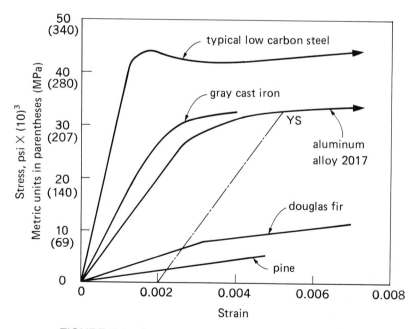

FIGURE 7.5 Stress–Strain Diagrams with Expanded Strain Scale

TABLE 7.1 Data from a Tension Test of a Steel Specimen

Initial Data: specimen = 1020 (low-carbon) steel; diameter = 0.505 in.; gage length = 2.000 in.

Lab Data		Calculated Results	
Load (lb)	Elong. (in.)	Stress (psi)	Strain (in./in.)
0	0	0	0
1 500	.0003	7 500	.000 15
2 500	.0006	12 500	.000 30
3 500	.0010	17 500	.000 50
4 500	.0013	22 500	.000 65
5 500	.0016	27 500	.000 80
6 500	.0020	32 500	.001 00
7 500	.0023	37 500	.001 15
8 500	.0026	42 500	.001 30
9 500	.0030	47 500	.001 50
10 000	.0032	50 000	.001 60
10 500	.0034	52 500	.001 70
11 000	.0036	55 000	.001 80
11 500	.0039	57 500	.001 95
12 000	.0043	60 000	.002 15
12 500	.0046	62 500	.002 30
13 000	.0051	65 000	.002 55
13 500	.0060	67 500	.003 00
14 000	.0073	70 000	.003 65
14 500	.0088	72 500	.004 40
15 000	.0118	75 000	.005 90

Final Data: ultimate load = 15 830 lb; diameter = 0.312 in.; gage length = 2.328 in.

CARBON STEELS

Terms such as *low-carbon* and *high-carbon* steels are used in this text, as in most technical texts and handbooks, when discussing engineering materials. There are two general classifications of steel: carbon steels and alloy steels. Carbon steels are those in which the principal ingredients are iron and carbon. Authorities vary slightly on the range of

CHAPTER SEVEN

low-, medium-, and high-carbon steel, but the following ranges are typical:

1. Low-carbon steel contains less than 0.30 percent carbon.
2. Medium-carbon or machinery steel has between 0.30 and 0.70 percent carbon.
3. High-carbon or tool steel has between 0.70 and 1.40 percent carbon.

PROPORTIONAL LIMIT

For most engineering materials, stress is proportional to strain up to the *proportional limit*. This limit is indicated on the graph in Figure 7.6 as the point where the stress–strain curve bends away from its straight section. Hooke's Law does not apply above this point. In other words, the formula relating stress to strain, $E = \sigma/\epsilon$, cannot be used above the proportional limit. For example, the initial straight sections of the curves all indicate that if the stress is doubled, the strain doubles. But the later sections of the curves clearly indicate that the strain can double or triple with little or no increase in stress.

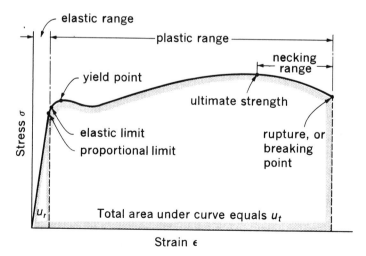

FIGURE 7.6 Stress–Strain Diagram for a Low-Carbon Steel

MODULUS OF ELASTICITY

Refer to Figure 7.6. The *modulus of elasticity E* is the measure of the slope of the initial straight part of the stress–strain curve. In Chapter 6, it was shown that $E = \sigma/\epsilon$; observe that E is a constant value up to the proportional limit. Above the proportional limit, the ratio of stress to strain is not constant.

ELASTIC LIMIT

Again, refer to Figure 7.6. Just above the proportional limit and close to it is a point labeled the *elastic limit*. This point divides the stress–strain curve into two parts, the elastic range and the plastic range. The engineering definition for *elasticity* is: the property that enables a material to return to its original shape and dimensions after being subjected to a static load and then having the load removed. This is a very important property. Almost all of the equations in this book are based on the assumption that stress is proportional to strain and that the stress at the elastic limit is not exceeded. Also, a machine part is expected to maintain certain dimensions during its usable life. For the engineering materials we will deal with, the elastic limit and the proportional limit are very close and therefore considered to be at the same point for practical problem solutions.

YIELD POINT

The *yield point* on the stress–strain curve is indicated in Figure 7.6. It is defined as the stress at that part of the curve where the strain increases with no increase in stress. In other words, the tangent to the curve at this point is horizontal. The ultimate stress (defined in the next section) also meets this definition but appears much farther along the curve. For low-carbon steels, there is usually a dip in the curve just past the yield point.

Normally, stress–strain diagrams are drawn for specimens loaded in tension. One reason is that ductile materials in compression have a tendency just to flatten out with no particular rupture (or breaking) point. Good examples of this are copper and gold, which can be rolled or beaten into thin sheets. In general, the yield points are the same for a material in compression or tension, although the ultimate tensile and compressive strengths may be different.

ULTIMATE TENSILE STRENGTH

The maximum stress that the material will take is referred to as its *ultimate strength*. This position on the stress–strain curve is shown in Figure 7.6. Note that in some materials the ultimate strength exceeds the breaking strength. The table of properties in the Appendix has a column headed "Ultimate Tensile Strength." Note that cast iron has an ultimate compressive stress much larger than its tensile strength. Some materials are so weak in tension that their major use is to support compressive loads, and handbooks will list ultimate compressive strengths. Such a material is concrete, whose ultimate tensile strength is only about 8 percent of its ultimate compressive strength.

NECKING RANGE

The necking range is indicated in Figure 7.6. Although a ductile material continually decreases in area as it elongates, the change in area is considered negligible up to the ultimate strength. Beyond the ultimate strength the cross-sectional area of a localized portion of the specimen decreases rapidly until rupture. Frequently, necking of the specimen can be observed. The part of the curve from the ultimate strength to the breaking (or rupture) point is called the *necking range*.

PLASTIC RANGE

In the *plastic range*, the stress is sufficient to cause a permanent deformation (or set) in the material. This permanent deformation can be noticed when a tensile test specimen that has been loaded to a stress in its plastic range is removed from the testing machine and measured. A large plastic range implies that a relatively large deformation occurs before rupture. Ductility and malleability are both measures of plasticity. These terms are defined next.

DUCTILITY

A *ductile material* is one that elongates a considerable amount before rupture. *Ductility* is a measure of the plasticity of a material when a tensile load is applied. It is measured by the change in length or the change in area and is usually expressed in percent. The percent length change equals final length minus original length, divided by original length, times 100:

$$\text{percent length change} = \frac{L_f - L_o}{L_o} \times 100$$

The percent area change is computed similarly, but these two percentages will not necessarily be the same. They are used for comparison with other length or area changes. Some examples of ductile metals are low-carbon steel, soft aluminum, copper, lead, and gold.

MALLEABILITY

Malleability is also a measure of the plasticity of a material, but the term refers to the amount a material will deform under a compressive load before rupture. The percentage change in area or length is a measure of malleability. Some authorities refer only to the very soft metals, such as copper, lead, silver, and gold, as being malleable.

BRITTLENESS

Brittleness is the opposite of plasticity and refers to a material that deforms very little before rupture. Cast iron, high-carbon steel, and glass are considered brittle. Do not confuse brittleness with stiffness, which is defined in the next section.

STIFFNESS

Stiffness is the property of being able to resist change in size and shape up to the elastic limit. Therefore, a stiffer material has a greater modulus of elasticity, which is really the measure of stiffness.

YIELD STRENGTH

You will have noticed that Figure 7.4 contains stress–strain diagrams for a selection of materials and that a number of these materials do not have an obvious yield point. In such a case, an arbitrary *yield strength* is determined from which design limits are based. The yield strength is determined by the *off-set method*. An example is shown for the aluminum curve in Figure 7.5. A point is marked off on the abscissa equal to 0.2 percent* of an inch, which is equivalent to a strain of 0.002. From this point, a line is constructed (the dashed line in Figure 7.5) parallel to the straight part of the curve. The stress value at the intersection of the curve and the constructed line is considered to be the yield strength.

HARDNESS

Hardness is the property of a material to resist wear and penetration. The earliest measure of hardness still in use is the *Mohs scale*. It was developed for minerals and is based on the fact that a harder material will "scratch" a softer one. The scale ranges between diamond (the hardest mineral) and talc (the softest mineral). Industry generally prefers systems based on the resistance to penetration. Some of the more common systems are the *Brinell*, the *Rockwell*, and the *Vickers*. Each of these systems is a variation of the principle that hard materials are more resistant to the penetration of a small steel ball or diamond point than are softer materials.

TRUE STRESS

The conventional stress–strain diagrams are based on calculations using the original length and original cross-sectional area of the specimen. These dimensions do change under load, and if, at each load reading, the calculations were based on the new length and new area, the diagram would be somewhat different. Below the elastic limit, the differences are minimal. All stress and strain data in this text are based on the conventional stress–strain diagrams, as are practically all handbook data.

*The 0.2 percent value is most common but may vary somewhat for different materials.

MODULUS OF RESILIENCE

The term *modulus of resilience* is used to denote the index of a material's ability to absorb energy up to the elastic limit. This energy is recoverable when the load is removed. The modulus of resilience is illustrated graphically by the area under the straight line part of the curve in Figure 7.6. The formula for determining the modulus of resilience is based on the area of a triangle.

$$\text{Area of triangle: } A = \frac{1}{2}(H \times B)$$

$$\text{Modulus of resilience: } u_r = \frac{1}{2}(\sigma \times \epsilon)$$

The units for the modulus of resilience are in.·lb/in.^{3} in the customary U.S. system and joule/metre^3 in the SI metric system. The stress and strain values are taken at the elastic limit. The modulus of resilience is useful when energy formulas are applied to such situations as impact on a machine or structural member. (You will be introduced to these applications in your succeeding courses.)

MODULUS OF TOUGHNESS

The *modulus of toughness* is an index of the work necessary to rupture a material and is represented graphically by the total area under the stress–strain curve (see Figure 7.6). An approximate value of the modulus of toughness can be obtained from the formula:

$$u_t = \frac{1}{3} \times \epsilon_r(\sigma_y + 2\sigma_u)$$

where
  u_t = modulus of toughness in in.·lb/in.^{3}
  σ_y = stress value at yield point
  σ_u = stress value at ultimate strength point of curve
  ϵ_r = strain value at rupture

Since work is defined in units of inch·pounds, these formulas provide information on the work required to deform a unit volume of material. Other methods of obtaining the modulus of toughness are by graphical calculus and Simpson's Rule. Toughness is important in applications where stresses in the plastic range of the material may be encountered, but where the resulting permanent set is unimportant

140   CHAPTER SEVEN

or even desirable. An example would be die-forming sheet steel for auto bodies.

The stress–strain diagrams indicate that low-carbon steels are much better than high-carbon steels from a toughness standpoint. On the other hand, high-carbon steels have a much greater modulus of resilience and are useful for springs and other shock load devices, which must maintain their original dimensions after the loads are removed.

COLDWORKING

The mechanical properties of a steel can be altered without changing the chemical composition. Steels may be ordered from the mill either *hot-rolled*, where the steel is heated before rolling to a desired shape, or *cold-finished*. The terms *cold-drawn* and *cold-rolled* are also used to indicate *coldworking*, which is the process of stressing a material to some point in its plastic range and then removing the load for the purpose of making a stronger material. Coldworking is usually accomplished at mills by drawing the material through dies or forcing it through rollers without prior heating of the material.

As an example, let us divide a piece of low-carbon hot-rolled steel and run a tensile test on one of the pieces. The typical stress–strain curve will be produced. The second piece is placed in the testing machine but is subjected only to some stress in the plastic range below the ultimate strength, as shown in Figure 7.7 and represented by point

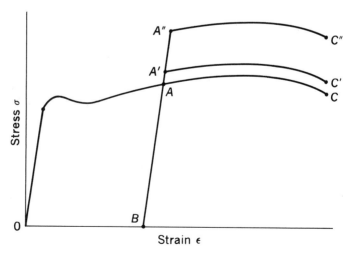

FIGURE 7.7 Stress–Strain Diagram of a Low-Carbon Steel Illustrating the Effect of Coldworking

A on the curve. The load is then removed from the specimen. As the load is removed, the specimen will backtrack along line A-B instead of the original curve. This indicates that the material has not come back to its original length as represented by the origin O, but has returned to some new length represented by the strain distance O-B. This difference in position between O and B is referred to as *permanent set* and is the result of coldworking the material.

If this specimen is remeasured and then loaded, a new stress–strain diagram will result, similar to the curve B-A"-C" in Figure 7.7. Note that the elastic limit has been raised, and the length of the plastic range has been shortened. The line A'-C' in Figure 7.7 indicates what the position of the curve would be if the original dimensions of the specimen had been used for the calculations in this second loading. The total strain of this coldworked specimen has been decreased approximately by the amount of permanent set.

FATIGUE

Failures of machine members that take repeated stresses are commonly due to *fatigue* of the metal. It has been found that rapid and repeated application of tensile and compressive forces on a test specimen may result in failure, even though the stress values may be well below the ultimate strength of the material. To run a fatigue test on a material, large numbers of test specimens are prepared. Each specimen has a successively lower stress placed on it, and the test is run until rupture. From these tests, a graph is plotted of stress versus the number of cycles of repeated stress. Such a graph is shown in Figure 7.8. The flat part of each curve is referred to as the *endurance limit*. The endurance limit for a specific material may vary, depending on how the load is applied. As an example, if there is not a complete reversal of stress but only applied and released tension, then the endurance limit will have a different value. Also, the endurance limit in torsion for ferrous metals may be only 50 percent of the limit obtained by the method of reverse bending.

The causes of fatigue failure have not been demonstrated experimentally. In the past, it was thought that the material changed its crystalline structure, and this caused failure. The present theory is that sudden changes in the shape of the machine member and inconsistencies in the material cause localized stresses far above the average stress in the material. These localized stresses are above the material's yield strength and cause permanent deformations to occur in these localities. Repeated permanent deformation in a small area eventually causes hairline cracks to develop. As the cracks enlarge, the area resisting the load continually decreases. This process contin-

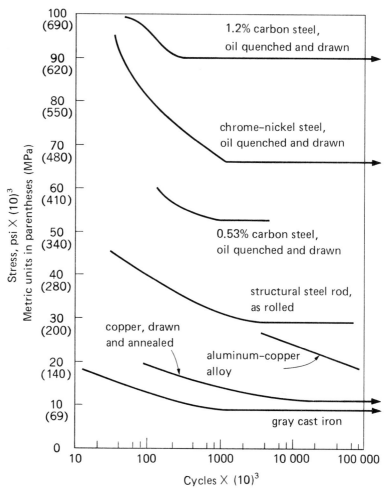

FIGURE 7.8 Stress versus Cycles to Failure from Endurance Tests

ues until the average stress on the resisting area exceeds the ultimate strength of the material, and the machine part suddenly fractures.

CREEP

Creep is defined as the continuing deformation of a material under constant stress for long periods of time. It has been found that metals and nonmetals creep at elevated temperatures, even at stresses below their yield points. How high is elevated? That depends on the material. Low-melting metals, such as lead, creep at room temperature. For steels, most handbooks do not list creep data below 750°F. The deformation due to creep is generally permanent. That is, if the high temperature and load were removed, very little of the deformation would be recovered.

It is obvious that for high-temperature applications, improperly designed steel members would soon deform to such an extent that failure would result. A typical problem would concern the design of a boiler shell with such a stress that the amount of deformation occurring in a period of 10 to 20 years would not be sufficient to cause failure. A standard considered satisfactory for many applications states that the strain should not exceed 1 percent (0.01 in./in.) in 100,000 hours (approximately 11 years of continuous use). This increase in strain is called *creep rate*, and it varies with stress, temperature, and time. Figure 7.9 indicates the creep curves for a material at a constant

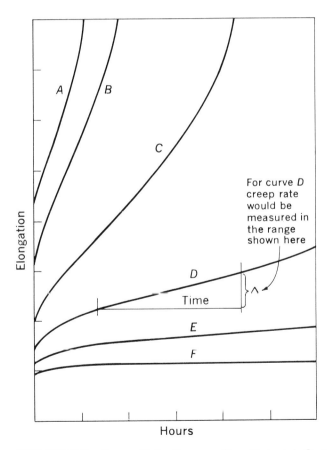

FIGURE 7.9 Creep Rate Curves: One Material, One Temperature, Various Stresses

temperature but with the applied stress successively lowered from curve A to F. Creep rate is usually measured along the flat part of the curve. Although no units are shown in Figure 7.9, the scale is such that curves A, B, and C would produce creep rates exceeding 1 percent, and curves D, E, and F would be at or below 1 percent.

LOW-TEMPERATURE IMPACT

Low temperature also affects materials. Some extremely northern sections of Canadian railroads, for example, cannot be run in the winter months because the weight of the trains splinters the rails. Although steel is not adversely affected from a strength or elasticity standpoint, the impact resistance does change.

Impact resistance is measured with the Charpy test apparatus. This device makes use of the fact that a friction-free pendulum swings as high on the up-swing side of its arc as the release position is on the other side of the arc. To perform a test, a specimen is inserted at the bottom of the arc, and the heavy pendulum is released. The impact bends or breaks the specimen. The pendulum's up-swing is shortened. This shortening of the up-swing provides data to determine the amount of energy absorbed by the specimen. A brittle material will snap easily and cause little change in the pendulum's up-swing.

For some steels, a relatively small temperature change (to $-20°$ or $-30°F$) can cause a ductile steel to behave like a brittle steel under impact conditions. Figure 7.10 relates Charpy impact test data to temperature. The ordinate is in units of ft·lb of energy that a specimen can absorb before breaking. Figure 7.10 shows that as the temperature rises, the specimen can absorb more energy. Steels for low-temperature work should have some nickel in their composition, because nickel changes the atomic structure of steel in such a manner that the conversion from a ductile to a brittle steel is delayed until a lower temperature is reached.

POISSON'S RATIO

The reduction in area of a tensile test specimen is apparent in the plastic range of the material, especially in the necking range. This reduction in area in a transverse direction also takes place in the elastic range and has a definite relation to the increase in length of the specimen. This reductive characteristic of solid matter was first noted by the French mathematician Siméon Poisson (1781–1840). If a bar of material is placed in tension along its longitudinal axis, the ratio of the lateral strain to the axial strain is known as *Poisson's Ratio*. The ratio is denoted by the Greek letter *mu*, μ.

Mechanical Properties of Materials 145

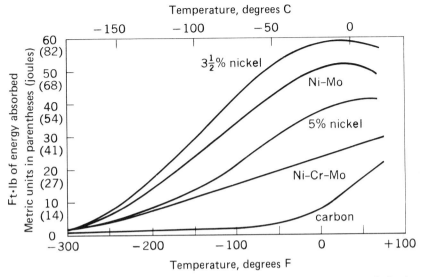

FIGURE 7.10 Charpy Impact Resistance of Annealed Carbon and Nickel-Alloy Steels at Low Temperatures

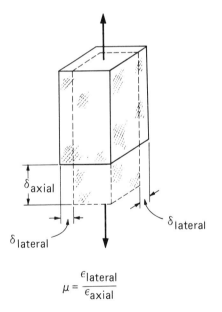

FIGURE 7.11 Lateral Strains Developing from Axial Loads and Poisson's Ratio

Figure 7.11 illustrates Poisson's Ratio. The formula is:

$$\mu = \frac{\epsilon_{lateral}}{\epsilon_{axial}}$$

It is of interest to note that there is no additional stress accompanying this effect (unless the lateral deformation is constrained). Poisson's Ratio is most commonly found to be between the values of 0.211 (for some cast irons) and 0.43 (for lead). Most steels have values between 0.283 and 0.292. These values indicate an increase in volume. If the volume should remain constant, as it does for rubber and most materials in the plastic range, Poisson's Ratio would be 0.50. This ratio can be applied to either tensile or compressive loads.

SUMMARY

This chapter deals with the mechanical properties of materials useful to engineers and technicians. The properties apply to materials in general, and to steel in particular.

The static tensile test is useful for determining proportional limit, modulus of elasticity, elastic limit, yield point or yield strength, ultimate strength, necking range, plastic range, ductility or malleability, brittleness, stiffness, modulus of resilience, the modulus of toughness, and permanent set.

Important characteristics obtained from other tests are hardness, true stress, endurance limit, creep, impact resistance, and Poisson's Ratio.

PROBLEMS

1. Refer to Figure 7.5 and approximate the modulus of elasticity for (a) pine and (b) steel.
2. For a given specimen of steel, the original diameter was 0.505 in., and the final diameter was 0.250 in. Determine the percent change in area.
3. An aluminum specimen had an original gage length of 2.00 in. and a final length of 2.86 in. Determine the percent change in length.
4. The following data were obtained from a tension test of 303 stainless steel. A standard ASTM specimen was used with a diameter of 0.505 in. and a gage length of 2.00 in.
    (a) Plot the stress–strain curve. Use a scale for the abscissa of about 0.05 in./in. to an inch. For the ordinate, use a scale of about 10 000 psi to a half inch.

(b) Plot the curve again for the first 0.1000 in./in. of strain, and expand the abscissa to a scale of 0.0010 in./in. to an inch.
(c) Determine the yield strength by the 0.2 percent off-set method.
(d) Determine by inspection or by calculation the ultimate strength, the rupture strength, the modulus of resilience, the modulus of toughness, and the percent change in the length.

Load (lb)	Elong. (in.)	Load (lb)	Elong. (in.)	Load (lb)	Elong. (in.)
0	0	10 250	.0086	12 820	.1025
3 200	.0004	10 250	.0090	13 000	.1120
5 450	.0012	10 420	.0096	13 900	.2000
6 825	.0020	10 490	.0112	16 100	.3500
8 050	.0028	11 000	.0170	17 050	.4500
8 850	.0040	11 200	.0340	18 000	.6000
9 300	.0048	11 530	.0445	18 450	.7500
9 725	.0062	11 720	.0555	18 800	.9000
10 000	.0072	12 000	.0662	19 050	1.0500
10 100	.0077	12 100	.0785	18 600	1.1500
9 875	.0081	12 650	.0885	15 050	1.2500

5. Plot a stress–strain curve for a material with a modulus of elasticity of 20 000 000 psi and an elastic limit, a proportional limit, and a rupture strength, all at 80 000 psi of stress. The strain at rupture is 0.010.
6. Which is the toughest material in Figure 7.4? Why? (Assume that brass ruptures at a strain of 0.30.)
7. From Figures 7.4 and 7.5, suggest an appropriate material for a leaf spring and one for a crane hook.
8. Determine (a) the modulus of resilience for aluminum in Figure 7.5, (b) the modulus of toughness for cast iron in Figure 7.5, and (c) the modulus of toughness for the lowest-carbon steel in Figure 7.4.

COMPUTER PROGRAM

The following program will perform your laboratory calculations when you run a static tensile test. You insert the load and gage readings, and the stress and strain will be calculated and presented in tabular form. The extensometer gage readings must be typed in just as you read them from the dial. Do not convert them to decimal inch. The program is arranged so that if your estimate of the required number of readings is low, two more readings can be taken each time that you indicate the test is unfinished.

```
10   REM   STRESSSTRAIN
20   PRINT "THIS PROGRAM WILL CONVERT TENSILE TEST DATA TO STRESS AND STRA
     IN."
21   PRINT "LOADS TO BE IN POUNDS AND THE GAGE LENGTH = 2 INCHES."
30   PRINT "REMEMBER TO TYPE IN THE LOAD FIRST AND THEN THE EXTENSOMETER G
     AGE"
31   PRINT "READING WHEN THE WORD 'DATA' IS DISPLAYED."
40   PRINT "TYPE IN THE ESTIMATED NUMBER OF READINGS YOU WILL BE TAKING."
50   INPUT N
60   PRINT "TYPE IN THE DIAMETER OF THE SPECIMEN."
70   INPUT D
80   LET A = 3.1416 * (D ^ 2) / 4
90   HOME
100  PRINT   TAB( 7)"LOAD"; TAB( 13)"GAGE"; TAB( 20)"STRESS"; TAB( 32)"STR
     AIN"
110  PRINT   TAB( 7)"LB"; TAB( 13)"READ"; TAB( 20)"PSI"; TAB( 32)"IN./IN."
120  FOR X = 1 TO N
130  PRINT "DATA"
140  INPUT P,G
150  LET S =  INT ((P / A) * 10 + .51) / 10
160  LET STRAIN = G / (2 * 10000)
170  PRINT  TAB( 7)P; TAB( 13)G; TAB( 20)S; TAB( 32)STRAIN
180  NEXT X
190  PRINT "IF FINISHED,TYPE 1, IF NOT TYPE 2."
200  INPUT T
210  IF T = 1 THEN   GOTO 240
220  IF T = 2 THEN   GOTO 300
240  END
300  LET N = 2
310  GOTO 120
```

8

Allowable Stress, Concentrated Stress, and Stress on Other Areas

OBJECTIVES

This chapter offers guidelines concerning safe stress values for practical applications. It demonstrates the use of handbook data to solve stress concentration problems involving notches and holes in material and includes an explanation of stresses developed on internal areas of a part other than the cross-sectional area. Upon completion of this chapter, you will have the necessary information for the design of safe structural members and machine parts (given known tensile or compressive forces exerted along one axis of the part) and will be able to implement this design under the direction of the chief design engineer.

INTRODUCTION

In our technical society today, it is most likely that the construction of a product or structure involves not only complicated plans but possibly hundreds of people who must communicate with one another. This chapter gives information that is directly related to design of structures and machines. You need to become familiar with the designations of standard steel shapes that are fabricated at the factory. You need to be able to determine how much stress can be placed on a structural member. Even if you do not become a designer, this knowledge is useful, because you must be alert for oversights in design, material, and construction. It is not uncommon for young engineers and technicians to catch an error in the plans or, during the construction process, to observe that someone is misreading or ignoring some detail in the plans.

STEEL BEAM DESIGNATIONS

Standard steel beams that can be ordered to size from a mill are generally specified by type of cross section, height of cross section, and weight per foot of length. Problem 2 at the end of this chapter specifies a C 3 × 6 beam, for example. The C designates a rolled channel steel beam (described in the Appendix). The first number signifies the overall height of the channel cross section in inches. The second number indicates the beam weighs 6 lb per foot of length. Other shape designations are S for standard I-beams, W for wide-flange beams, and L for angle beams. The Appendix gives specifications for a number of shapes. S and W shapes have designations similar to channel beams; that is, an S 10 × 35 indicates a standard beam with a cross section 10 in. high and a weight of 35 lb per foot of length. Angles are specified somewhat differently. An L 6 × 4 × 1/2 refers to an angle with legs 6 in. by 4 in. and a metal thickness of 1/2 in.

ALLOWABLE STRESS

In the design of machines or structures, individual parts should not become too distorted to perform their functions. Designers usually agree that the limit of usable strength for a ductile material (elongations of over 5 percent) is represented by the elastic limit; however, the yield point is generally used in calculations. The limit of usefulness for brittle materials (elongations under 5 percent) is considered to be the ultimate strength.

Let us consider a machine member of ductile steel. As designers, we are required to specify a suitably sized member to hold a particular load. We would not design the member to be stressed up to its elastic limit for several reasons:

1. Although a load may have been specified for the part, our experience as a designer will tell us whether or not there is a possibility of an overload for this application.
2. The yield stress, modulus of elasticity, and other characteristics are average values obtained from a large number of specimens. The specific piece chosen for our machine member may fall below the average value.
3. There may be slight imperfections in the material that would lower its yield or ultimate strength. This is particularly true of timber with knots that interfere with stress distribution.

Civil engineers and technicians have codes that provide allowable stresses for various types of structural steel. For example, A36 steel is a common structural steel. It has a yield point of 36 000 psi, and the American Institute of Steel Construction (AISC) states that the allowable static tensile or compressive stress is 22 000 psi. A sample of the AISC code with allowable stress information is given in the Appendix. The term *allowable stress*, sometimes called *working stress*, means that a structural member should be designed to take that particular stress value. The various engineering societies offer guides to the designer.

A guide for the machine designer is the *factor of safety*. For example, a factor of safety, based on the yield point, of 2 means that the allowable stress is determined by dividing the yield stress by 2. Thus,

$$\sigma_{allowable} = \frac{\text{yield stress}}{\text{factor of safety}}$$

The yield stress in the preceding formula may be replaced by the ultimate stress or the endurance limit, depending on the various material or loading conditions. Typical recommendations for the factors of safety are given in Table 8.1.

For the purpose of solving problems in this chapter, first determine whether or not an allowable stress value is given in the Appendix. If not, use Table 8.1, and choose an appropriate safety factor. Don't worry if you choose a factor of safety of 5 for cast iron and another student chooses 7. This is a matter of judgment based on experience; in actual practice, it will be decided by the chief engineer. The following metric sample problem illustrates the use of allowable stress and the factor of safety.

TABLE 8.1 Factor of Safety

Load Condition	For Steel and Ductile Metals and Based on Yield Point	For Cast Iron and Brittle Metals and Based on Ultimate Strength
Static load	1.5–2	5–7
Mild shock (gradual load change)	3	7–8
Shock	5–7	15–20
Fatigue load (based on endurance limit)	2.5	2.5

METRIC SAMPLE PROBLEM 8.1
Allowable Stress

Problem

A C1020 cold-rolled steel support block has cross section dimensions of 58 mm by 72 mm. What safe static load can the block support? *Note*: Stress is in pascals, which is the same as newton per square metre (N/m²). Therefore, dimensions must be in metres:

58 mm = 0.058 m

72 mm = 0.072 m

Solution

From the Appendix, the yield strength of C1020 CR is 441 MPa. From Table 8.1, we will choose a safety factor of 1.5.

$$\sigma_a = \frac{YS}{1.5} = \frac{441}{1.5} = 294 \text{ MPa}$$

$$F = \sigma_a \cdot A = 294(10)^6 \cdot (0.058 \cdot 0.072) = 1.23(10)^6 \text{ N} = 1.23 \text{ MN}$$

Thus, the load in kilograms is:

$$\frac{1.23(10)^6}{9.81} = 0.125(10)^6 \text{ kg} = 125\,000 \text{ kg} = 125 \text{ Mg}$$

CONCENTRATED STRESS

Figure 8.1(a) represents front and side views (FBD) of a steel bar under a tensile load. The bar is of constant depth but varying width. The dotted lines represent lines of stress flow. Think of the stress flow lines as rubber bands holding the top and bottom surfaces together. When the stress flow lines are evenly spaced, the stress is evenly distributed over the cross section of the material. When the lines are close together, the stress is greater than when the lines are spaced further apart. Notice that the stress is more concentrated where the lines flow around the notch and around the hole. Figure 8.1(b) is an FBD showing the stress distribution on the critical area A-A.

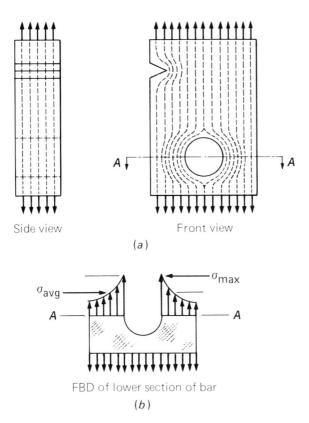

FIGURE 8.1 Stress Patterns

Because the mathematics necessary for determining the magnitude of the concentrated stress is generally too involved for most applications, experimental methods have been developed. Photoelastic experiments are the most common. Information from these experiments is presented in the form of graphs similar to the one in Figure 8.2. From the graph, the designer chooses the appropriate stress concentration factor K and multiplies it by the average stress to obtain the maximum stress appearing on the cross section. Expressed as a formula:

$$\sigma_{max} = K \cdot \sigma_{avg}$$

Sample Problem 8.2 illustrates a concentrated stress situation.

154 CHAPTER EIGHT

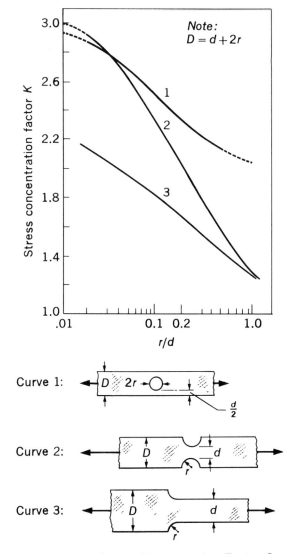

FIGURE 8.2 Stress Concentration Factor Curves

SAMPLE PROBLEM 8.2
Concentrated Stress

Problem

What is the maximum allowable load that can be applied to the flat plate in the accompanying sketch? Assume that the allowable stress =

15 000 psi; D = 1½ in.; d = 3/4 in., r = 3/8 in.; and plate thickness = 1/2 in.

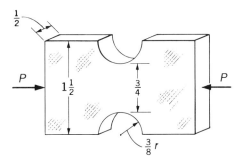

Solution

$$\frac{r}{d} = \frac{3/8}{3/4} = 0.5$$

From curve 2, Figure 8.2, K = 1.58. Therefore:

$$\sigma_{avg} = \frac{\sigma_{max}}{K} = \frac{15\ 000}{1.58} = 9500 \text{ psi}$$
$$P = \sigma_{avg} \cdot A_d = 9500 \times 3/4 \times 1/2 = 3565 \text{ lb}$$

STRESS INDUCED ON OTHER AREAS DUE TO A TWO-FORCE SYSTEM

It has been stated previously that, if a specimen is placed in tension, the critical area resisting the tensile load is perpendicular to the direction in which the force is applied. Now let us examine a specimen and imagine that it is cut in the middle at a 45° angle and that the two halves are then placed together again with a free-fitting dovetail joint. It can be demonstrated readily with such a model that if tension is applied, as in Figure 8.3(*a*), the two pieces will slide apart. This is an indication that shear forces are in evidence.

Figure 8.3(*b*) is an FBD of the left-hand piece. From mechanics, we know that a force can be broken down into two or more components. It is of importance to us at this stage to break the force F into rectangular components, one parallel to the inclined plane and the other

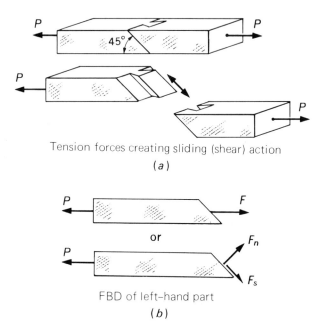

FIGURE 8.3 Forces Acting on Inclined Areas

perpendicular (normal) to the plane. Note that the force component F_s is a shear force.

The trigonometric formulas for obtaining shear stress, τ, and normal stress, σ_n, (tensile or compressive) are as follows:

$$\tau = \frac{F}{2A} (\sin 2\theta)$$

$$\sigma_n = \frac{F}{A} (\cos^2 \theta)$$

where

F = internal force on cross-sectional area
A = cross-sectional area
θ = angle between cross-sectional area and inclined area

These formulas are derived later in Sample Problem 8.3.

Interestingly enough, if the stresses in tension and shear are calculated for areas taken at ever-increasing angles to the cross section until rotation through 180° is accomplished, we will find that these stress values plot a circle. This relationship was originally developed by the German engineer Karl Culmann (1821–1881) and generalized by Otto Mohr (1835–1918). The circle is referred to as Mohr's Circle (see Sample Problem 8.3).

Note that for the type of loading we are dealing with in Figure 8.3 and in Sample Problem 8.3, the maximum shear stress is just half of the maximum tensile stress. The maximum shear stress appears on an area inclined at an angle of 45° to the cross-sectional area of the structural member. Also note that the angle θ, which the inclined area makes with the cross-sectional area, is exactly half of the corresponding central angle inscribed in Mohr's Circle. This angular relationship is true for all situations in which Mohr's Circle is used.

Procedure for Sketching Mohr's Circle

This procedure is limited to problems concerning tensile or compressive two-force systems. The steps will be modified somewhat for the combined stress problems in Chapter 20.

Step 1 Sketch a circle of convenient size.

Step 2 A vertical axis representing shear stress is drawn tangential to the circle. Put the circle to the right of the vertical axis if tension is applied to the cross section and to the left of the vertical axis if compression is applied. The horizontal axis represents tensile (or compressive) stress and is drawn through the center of the circle. By agreement among engineers, σ_t is shown as a plus value, and σ_c is shown as a minus value in the circle (see Mohr's Circle in Sample Problem 8.3).

Step 3 If the tensile (or compressive) stress on a cross-sectional area is known, it may be placed on the circle at the horizontal axis (see Sample Problem 8.3). This may be done because no shear stress has been applied to the cross section, and therefore no shear stress is developed. Any point on the circle represents the shear and tensile (or compressive) stresses on some area.

Step 4 The angle that the radius to a point on the circle makes with the horizontal is exactly twice the angle the inclined area makes with the cross-sectional area.

Step 5 If a stress (tensile, compressive, or shear) is given for some area at a given angle to the cross-sectional area, then the radius of the circle can be determined and the circle completed.

Step 6 The center of Mohr's Circle is always on the horizontal axis.

158 CHAPTER EIGHT

SAMPLE PROBLEM 8.3
Stress Induced on Other Planes

Problem

(a) Develop the general formulas for σ_n and τ on an area at any angle to the line-of-action of a two-force system. (b) Show that the stress values of at least three sectional areas form coordinates of points on a circle. (c) Determine θ for the areas that have maximum and minimum tensile and shear stresses.

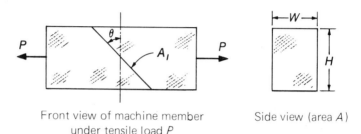

Front view of machine member under tensile load P

Side view (area A)

Symbols

P = load applied to part
A = area of cross section, width W, height H
A_I = area of section inclined to cross section
θ = angle between A and A_I
F = force on cross section, opposing P
F_n = force component normal to inclined area
F_s = force component parallel to inclined area
σ_n = stress normal to A_I
τ = shear stress

Solution (a)

Draw the FBD of the left half with F broken down into components.

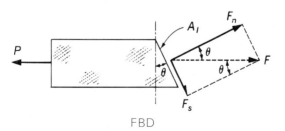

FBD

From the trigonometric relations in the FBD, we have:

$$A_l = \frac{A}{\cos \theta}$$

$$F_n = F \cdot \cos \theta$$

$$F_s = F \cdot \sin \theta$$

$$\sigma_n = \frac{F_n}{A_l} = \frac{F \cdot \cos \theta}{A/\cos \theta} = \frac{F}{A}(\cos^2 \theta)$$

$$\tau = \frac{F_s}{A_l} = \frac{F \cdot \sin \theta}{A/\cos \theta} = \frac{F}{A}(\sin \theta)(\cos \theta) = \frac{F}{2A}(\sin 2\theta)$$

Note: We have obtained σ_n and τ in terms of F and A, which eliminates the necessity of computing F_n, F_s, and A_l.

Solution (b)

We will choose the cross section and areas where angle $\theta = 15°$ and $40°$. Let $P = 800$ lb, $W = 1.5$ in., $H = 2$ in. For the cross section ($\theta = 0$):

$$\sigma_n = \frac{F}{A} = \frac{800}{1.5 \cdot 2} = 267 \text{ psi}$$

$$\tau = 0$$

For the section at $\theta = 15°$:

$$\sigma_n = \frac{F}{A}(\cos^2 \theta) = 267 (0.966)^2 = 250 \text{ psi}$$

$$\tau = \frac{F}{2A}(\sin 2\theta) = \frac{800}{2 \cdot 3}(0.5) = 67 \text{ psi}$$

For the section at $\theta = 40°$:

$$\sigma_n = 267 (0.766)^2 = 157 \text{ psi}$$

$$\tau = \frac{267}{2}(0.985) = 131 \text{ psi}$$

For the purpose of drawing a graph, all tensile stresses will have a plus sign, and compressive stresses will have a minus sign; shear stresses directed down and to the right will have a plus sign.

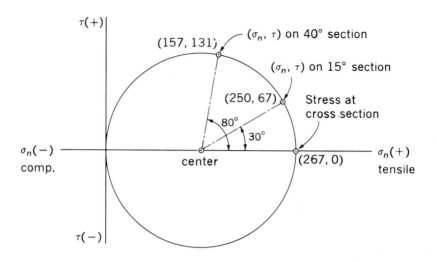

Solution (c)

From Mohr's Circle, we can determine that the maximum σ_n (267 psi) appears on the cross section; minimum σ_n (0 psi) appears on a section of $\theta = 90°$ from the cross section (180° on the circle); there are no compressive stresses; maximum τ (133 psi) appears on a section of $\theta = 45°$ from the cross section (90° on the circle); and minimum τ (0 psi) appears on the cross section and $\theta = 90°$ from the cross section. *Note:* Refer to engineering graphics texts for the procedure used in constructing a circle when you know three points on its circumference.

SUMMARY

Steel beams, which are rolled to various shapes at the factory, have specific designations for easy identification.

Allowable stress, sometimes called design stress or working stress, is set by codes for structures and by recommended factors of safety for machine parts. The formula for allowable stress is:

$$\sigma_{\text{allowable}} = \frac{\text{yield stress}}{\text{factor of safety}}$$

The yield stress may be replaced by ultimate stress or endurance limit, depending on circumstances.

The formula for concentrated stress is:

$$\sigma_{max} = K \cdot \sigma_{avg}$$

K is the concentrated stress factor taken from an appropriate graph.

A two-force system induces stresses on areas at angles other than 90° to the system. The normal stress (tensile or compressive) on any area is:

$$\sigma_n = \frac{F}{A}(\cos^2 \theta)$$

The shear stress (τ) on any area is:

$$\tau = \frac{F}{2A}(\sin 2\theta)$$

where
  θ = angle between cross-sectional and inclined areas
  F = force on cross section
  A = area of cross section

Data concerning allowable stress, factor of safety, and concentrated stress may be obtained from appropriate handbooks. On occasion, various engineering societies are referred to by the abbreviations listed below:

1. ASCE, American Society of Civil Engineers
2. ASME, American Society of Mechanical Engineers
3. AISC, American Institute of Steel Construction
4. AISI, American Iron and Steel Institute
5. ASTM, American Society for Testing and Materials

Allowable Stress

For reference, see the material strengths and various codes giving allowable stresses listed in the Appendix. Safety factors are shown in Table 8.1.

1. Determine the minimum cross-sectional area for a steel punch. Use AISI C1040 steel, and refer to the safety factor table. The punch is to take a 100 000 lb shock loading.

2. A C 3 × 6 channel made from ASTM A36 steel is used as a tension member in a structure. According to the AISC code, what allowable tensile load can it carry?
3. A 4 × 4 hard brass post is placed on a pine sill, as shown in Figure 8.4. What allowable compressive static load P can be applied to this assembly?
4. A W 8 × 31 structural shape is cut to a 15 in. length and placed on end. It is used to support a compressive load of 130 000 lb. What is the factor of safety based on the yield point of ASTM A36 steel?
5. Using A36 steel, design a square connector pin and support rods (refer to Figure 8.5) to resist a 55 000 lb. tensile load. Consider tension in the support rods and compression and shear in the pin, but do *not* consider stress concentration. For compression, use the allowable axial compressive stress.
6. (SI Units) A short aluminum rod has a diameter of 9 mm. Using a safety factor of 3 based on the yield stress, determine the safe tensile or compressive load that may be applied. Give the answer in kilograms.

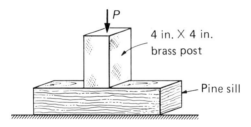

FIGURE 8.4 Problem 3

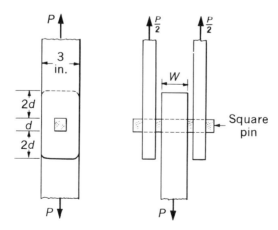

FIGURE 8.5 Problem 5

Allowable, Concentrated, and Other Stresses   163

Concentrated Stress

The graph of concentrated stress is shown in Figure 8.2.

7. Refer to Sample Problem 8.2. Solve for the allowable load if the grooves have a radius of 1/2 in.
8. Consider the flat plate for curve 3 of Figure 8.2 to be a connecting member in a structure with a tensile load of 15 000 lbs. Does the member meet AISC specifications for A36 steel? Assume that $D = 4$ in., $d = 3$ in., $r = 1/2$ in., and plate thickness $= 3/8$ in.
9. Consider the plate for curve 1 of Figure 8.2 to be an FBD of a flat bar tension member. The bar is hard brass. Use a safety factor of 2 based on the yield point, and determine the load that can be applied to the net cross-sectional area at the 1/2 in. diameter hole. Assume that $D = 3\frac{1}{2}$ in. and plate thickness $= 1/4$ in.

Stress on Other Areas

10. Refer to Sample Problem 8.3. Use the formulas to calculate the stresses on areas at 30° and 60° to the cross section, and compare your answers to those obtained by using Mohr's Circle.
11. A 45 000 lb tensile force is applied to a rectangular solid with a cross section 3 in. high and 2 in. wide. Using Mohr's Circle, determine the shear and tensile stresses on the cross section and on sections cut successively 9° apart (limit: 45° from the cross section).
12. Two wood blocks are glued together as in the sketch in Sample Problem 8.3. They are 3 in. wide and 5 in. high. What load will cause a shear stress of 400 psi in the glue (a) when the angle θ is 30° and (b) when the angle θ is 45°?
13. (Extra Credit) In Figure 8.6, a 4 × 4 timber post has its grain running at a 15° angle to the longitudinal axis of the post. Using

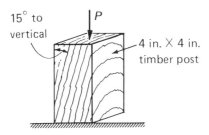

FIGURE 8.6   Problem 13

Mohr's Circle, determine the allowable compressive load. Refer to the Appendix for allowable stresses.

14. (SI Units)   Refer to Sample Problem 8.3. If $P = 70$ kN and $\theta = 30°$, determine the shear stress and normal stress to the inclined area. The cross-sectional area of the material is 50 mm wide and 200 mm high.

COMPUTER PROGRAM

The following program will calculate the normal and shear stresses on a plane at any angle to the cross section. If you want to plot Mohr's Circle, you can "walk" the program through a number of varying angles and obtain the X and Y coordinates (normal stress and shear stress) for different points on the circle.

```
10   REM "MOHRSCIRCLE"
15   HOME
20   PRINT "THIS PROGRAM WILL CALCULATE THE NORMAL STRESS AND SHEAR STRESS
     ON A PLANE AT ANY ANGLE TO THE CROSS SECTION."
21   PRINT " THE UNITS ARE TO BE POUNDS, SQUARE INCHES, AND DEGREES."
22   PRINT "TYPE IN THE APPLIED FORCE, THE CROSS SECTIONAL AREA, AND THE A
     NGLE OF INCLINE TO THE CROSS SECTION."
30   INPUT F,A,AA
40   LET Z = 3.1416 / 180
50   LET SN = INT ((F / A) * ( COS (AA * Z)) ^ 2 * 10 + .51) / 10
60   LET SS = INT ((F / (2 * A)) * ( SIN (2 * AA * Z)) * 10 + .51) / 10
70   PRINT
80   PRINT "NORMAL STRESS             ="; TAB( 28)SN;" PSI."
81   PRINT "SHEAR STRESS              ="; TAB( 28)SS;" PSI."
82   PRINT "ANGLE FROM CROSS SECTION ="; TAB( 28)AA;" DEG."
90   PRINT
100  PRINT "IF YOU WANT TO TRY ANOTHER INCLINED AREA, TYPE 1, IF NOT TYPE
     2."
110  INPUT C
120  IF C = 1 THEN  GOTO 150
130  IF C = 2 THEN  GOTO 210
140  IF C < 1 OR C > 2 THEN  GOTO 190
150  PRINT "TYPE IN THE ANGLE OF INCLINE TO THE CROSS SECTION."
160  INPUT AB
170  LET AA = AB
180  GOTO 40
190  PRINT "DOES NOT COMPUTE. TYPE 1 OR 2."
200  GOTO 110
210  PRINT "END OF PROGRAM.  TYPE 'RUN' FOR ANOTHER PROBLEM."
220  END
```

9

Statically Indeterminate Machine and Structural Members

OBJECTIVES

This chapter will help you to recognize situations that cannot be handled by statics alone. It will also help you to (1) determine the strain relationships between two materials supporting an axial load, (2) determine the load on each material, and (3) solve for loads caused by temperature changes.

INTRODUCTION

The term *statically indeterminate* indicates that there are situations in which knowledge of statics alone cannot provide the answer to the question, "What force is acting on the machine member?" However, there may be other ways to obtain an answer. In our discussion, a knowledge of stress–strain relationships will help. Basically, a knowledge of strain is used to determine stress. Problems concerning forces resulting from temperature changes are closely related to the above situations.

STATICALLY INDETERMINATE MACHINE MEMBERS

There are situations when the equations of statics ($\Sigma F_x = 0$, $\Sigma F_y = 0$, and $\Sigma M = 0$) are not sufficient to solve for the forces applied to the member of a machine or structure. As an example, look at Figure 9.1.

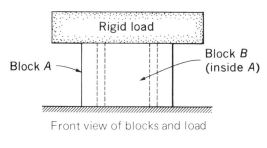

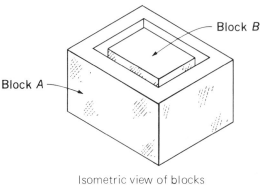

FIGURE 9.1 Statically Indeterminate Supports

Both outside and inside support blocks are steel and have the same horizontal cross-sectional areas. Our first thought might be that each block supports half of the total load because the areas resisting the load are equal. Intuitively we are making the assumption that the stresses are the same, and because both blocks have the same areas, the loads are the same. Although the answer is correct in this instance, we can run into trouble if we fail to make allowances for strain. If block B were brass, can we still assume each block takes half the load? No. Strains are the same, but the materials are different; therefore, the stresses will be different. The stress–strain relationship is the key to solution of problems of this type.

Previously, the formulas $\sigma = F/A$, $E = \sigma/\epsilon$, and $\delta = \epsilon \cdot L$ have been used. They are now combined to form a more useful formula relating deformation to the force applied and the material resisting:

$$\delta = \frac{F \cdot L}{A \cdot E}$$

which is read: *deformation equals force times length divided by area times modulus of elasticity.* The derivation is left as an exercise (see Problem 1 at the end of this chapter).

Procedure for Solving Statically Indeterminate Problems

Step 1 Draw an FBD of the complete assembly and of the members as needed.

Step 2 Determine the relationship of the deformation of one member to the other. Ask yourself if the deformations are equal to each other or whether they must be added or subtracted to obtain the total deformation of the complete assembly. This is a very important step. The appropriate deformation equations are as follows:

$$\delta_a = \delta_b$$
$$\delta_a = \delta_b \pm \text{constant}$$
$$\delta_{\text{total}} = \delta_a \pm \delta_b$$

where
δ_a = deformation of material a
δ_b = deformation of material b

For the case where two members are sharing a known load and the deformation is not known, adjust to one unknown force as follows: Suppose the total load is 150 lb. Then the sum of the forces on each member must equal 150 lb:

$$150 \text{ lb} = F_a + F_b$$

One unknown force can now be written in terms of the other:

$$F_a = 150 \text{ lb} - F_b$$

Step 3 Substitute FL/AE for each of the deformations and proceed with the solution.

Sample Problems 9.1, 9.2, and 9.3 illustrate typical situations.

SAMPLE PROBLEM 9.1
Statically Indeterminate Machine Members with Axial Loads

Problem

A hollow steel cylinder 0.25 in. thick, with a 12 in. outside diameter, and 10 in. long is filled with concrete. It is used as a post to support a 1000 kip load (1 000 000 lb). Given $E_{con} = 4.5(10)^6$ psi, determine (a) the load each material supports and (b) the length of the post under load.

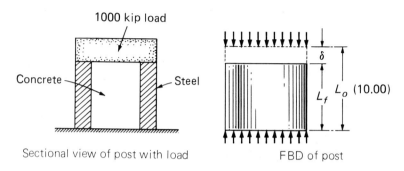

Sectional view of post with load — FBD of post

Solution (a)

Step 1 Note that the FBD is the first step for both solutions (a) and (b).

Step 2 The rigid load deforms both materials in equal amount:

$$\delta_{con} = \delta_{st}$$
$$F_{con} + F_{st} = 1\,000\,000 \text{ lb}$$

Step 3

$$\frac{F_c L_o}{A_c E_c} = \frac{F_s L_o}{A_s E_s}$$

$$\frac{F_c L_o}{A_c E_c} = \frac{(1\,000\,000 - F_c)L_o}{A_s E_s}$$

Divide out the L_o's:

$$\frac{F_c}{104(4.5 \times 10^6)} = \frac{1\,000\,000 - F_c}{9(30 \times 10^6)}$$

Multiply through by the denominator of the first term to find the answer:

$$F_c = (1\,000\,000 - F_c) \times 1.73 = 1\,730\,000 - 1.73 F_c = 634\,000 \text{ lb}$$
$$F_s = 1\,000\,000 - F_c = 366\,000 \text{ lb}$$

Solution (b)

$$\delta_{con} = \delta_{st} = \frac{F_c L_o}{A_c E_c} = \frac{634\,000 \times 10}{104(4.5 \times 10^6)} = 1350(10)^{-5} = 0.0135 \text{ in.}$$

$$L_f = 10.00 - 0.0135 = 9.99 \text{ in. (rounded)}$$

SAMPLE PROBLEM 9.2
Statically Indeterminate Problem

Problem

The following sketch indicates that, before loading, the brass wire is exactly 40 ft long and the steel wire is 40.05 ft long. Each wire has a cross-sectional area of 0.25 in.^{2}. When the wires pick up the full 10 kips, what load does each wire carry? (The load is large enough to strain both wires.)

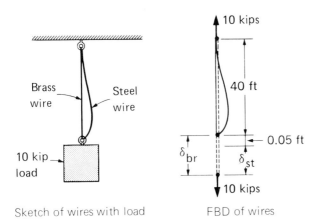

Sketch of wires with load   FBD of wires

Solution

Step 1 Draw the FBD.

Step 2 We observe from the FBD that the δ relationship is:

$$\delta_{br} = \delta_{st} + 0.05 \text{ ft}$$
$$10 \text{ kips} = F_{br} + F_{st}$$

Step 3

$$\frac{F_{br}L_{br}}{A_{br}E_{br}} = \frac{F_{st}L_{st}}{A_{st}E_{st}} + (0.05 \text{ ft} \times 12 \text{ in./ft})$$

$$\frac{F_{br}(40 \times 12)}{0.25(16 \times 10^6)} = \frac{(10 \text{ kips} - F_{br})(40.05 \times 12)}{0.25(30 \times 10^6)} + 0.6 \text{ in.}$$

Multiply through by $0.25(16 \times 10^6)(30)$:

$$F_{br}(30)(40 \times 12) = (10 \text{ kips} - F_{br})(40.05 \times 12)(16)$$
$$+ 0.6(0.25)(16 \times 10^6)(30)$$

$$14\,400 F_{br} = (10 \text{ kips})(7690) - 7690 F_{br} + 72(10)^6$$

$$22\,000 F_{br} = 76.90(10)^6 + 72(10)^6$$
$$F_{br} = 6740 \text{ lb}$$
$$F_{st} = 10\,000 \text{ lb} - F_{br} = 3260 \text{ lb}$$

Alternate Solution

Step 2 If we view the problem such that the load is applied gradually, it is apparent that the brass wire takes all of the load up to the point where its deformation is 0.05 ft. From this point, the steel wire takes a share of the load, and $\delta_{br} = \delta_{st}$.

Step 3 First, we find the force required to stretch the brass 0.05 ft.

$$\delta_{br} = 0.05 \text{ ft} = \frac{F_{br} L_{br}}{A_{br} E_{br}}$$

Rearranging, we have:

$$F_{br} = \frac{(0.05 \times 12 \text{ in./ft})(A_{br})(E_{br})}{L_{br}}$$

$$= \frac{(0.05 \times 12)(0.25)(16 \times 10^6)}{40 \times 12} = 5000 \text{ lb}$$

The remaining load to be shared is 10 000 lb minus 5000 lb, or 5000 lb; and for the remaining load:

$$\delta_{br} = \delta_{st}$$
$$\frac{F_{br} L_{br}}{A_{br} E_{br}} = \frac{(5000 - F_{br}) L_{st}}{A_{st} - E_{st}}$$
$$\frac{F_{br}(40.05 \times 12)}{0.25(16 \times 10^6)} = \frac{(5000 - F_{br})(40.05 \times 12)}{0.25(30 \times 10^6)}$$
$$30 F_{br} = 5000(16) - 16 F_{br}$$
$$F_{br} = \frac{5000(16)}{46} = 1740 \text{ lb}$$

Brass takes its initial load (5000 lb) plus 1740 lb:

$$F_{br} = 6740 \text{ lb}$$

Steel takes 5000 lb minus 1740 lb:

$$F_{st} = 3260 \text{ lb}$$

Statically Indeterminate Machine and Structural Members   171

SAMPLE PROBLEM 9.3
Statically Indeterminate Problem

Problem

A clamp consists of a special 2 in. diameter steel bolt with a minor thread area of 2.3 in.^{2}, 12 threads per inch, and two rigid steel plates. Between the plates, two aluminum prisms, each 3 in.^{2} in area, are placed. Neglect any deformation in the plates, nut, and bolt head. Determine (a) what stress is placed on the bolt and aluminum prisms when the nut is turned down 1/4 turn and (b) what deformation occurs in the bolt and prisms.

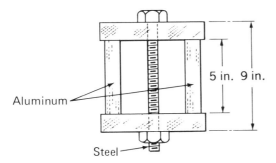

Solution

If the aluminum did not deform, then the bolt would take the total strain; however, the bolt will extend some, and the aluminum prisms will compress.

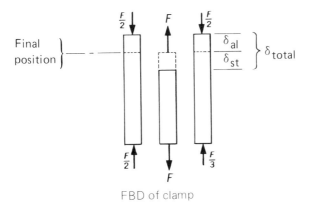

FBD of clamp

Step 1 Consider the situation as shown in the FBD.

Step 2

$$\delta_{total} = \delta_{st} + \delta_{al}$$
$$F_{st} = F_{al}$$
$$\delta_{total} = \frac{1}{4} \times \frac{1}{12} = \frac{1}{48} = 0.0208 \text{ in.}$$

Step 3

$$0.0208 = \frac{F_{st}L_{st}}{A_{st}E_{st}} + \frac{F_{al}L_{al}}{A_{al}E_{al}} = \frac{F_{st}(9)}{2.3(30 \times 10^6)} + \frac{F_{st}(5)}{6(10 \times 10^6)}$$

total area of aluminum ⬆

$$0.0208(10)^6 = 0.1305 F_{st} + 0.0834 F_{st}$$
$$F_{st} = F_{al} = 97\,400 \text{ lb}$$

Answer (a)

$$\sigma_{st} = \frac{97\,400}{2.3} = 42\,300 \text{ psi T}$$

$$\sigma_{al} = \frac{97\,400}{6} = 16\,200 \text{ psi C}$$

Answer (b)

$$\delta_{st} = \frac{42\,300 \times 9}{30 \times 10^6} = 0.0127 \text{ in.}$$

$$\delta_{al} = \frac{16\,200 \times 5}{10 \times 10^6} = 0.0081 \text{ in.}$$

EFFECTS OF TEMPERATURE CHANGE

Most materials expand (in all directions) when their temperature is raised and contract when their temperature is lowered. If the material is unrestricted in its movement, there is no change in the stress on the material, although strain takes place. Stresses are set up in the material if its movement is restricted. The change in each inch of length of a member, per degree Fahrenheit, is called the *linear coefficient of thermal expansion*. The symbol is the Greek letter *alpha,* α. The Appendix contains a table of coefficients for some common materials. The units of the coefficient are inch/inch/degree Fahrenheit. In other words, the coefficient is strain per degree Fahrenheit. To determine the strain produced by some temperature change, simply multiply the coefficient by the temperature change:

$$\epsilon = \alpha(T_f - T_i) = \alpha \cdot \Delta_t$$

where
T_f = final temperature in degrees Fahrenheit
T_i = initial temperature in degrees Fahrenheit
Δ_t = change in temperature

When the total deformation of a member is required, we simply multiply the strain by the original length and obtain:

$$\delta = \alpha \cdot \Delta_t \cdot L$$

which is read: *deformation equals coefficient of thermal expansion times change in temperature times original length.*

Procedure for Solving Temperature Problems

Step 1 First, draw FBDs as necessary. Then, find the strain, or deformation, resulting from the temperature change that each member would have if unrestrained.

Step 2 Consider the unrestrained length as the original length, and continue solution as outlined in the first section of this chapter. Remember that the deformation is the difference between the unrestrained and restrained lengths.

Sample Problems 9.4, 9.5, and 9.6 and metric Sample Problem 9.7 deal with the effects of temperature change.

SAMPLE PROBLEM 9.4
Structural Members Subjected to Temperature Change

Problem

A 4 in.^{2} steel bar, 5 in. long, just touches two rigid walls at 70°F. How much force is exerted on the bar when its temperature is raised to 150°F?

Solution

Step 1 If unrestrained, the deformation of the bar due to temperature is

$$\delta = \alpha \cdot \Delta_t \cdot L = (6.5 \times 10^{-6})(150° - 70°)(5 \text{ in.}) = 0.0026 \text{ in.}$$

Step 2 We now consider the bar to be squeezed from 5.0026 in. to 5.0000 in.

$$\delta = \frac{FL}{AE}$$

$$F = \frac{\delta AE}{L} = \frac{0.0026(4)(30 \times 10^6)}{5} = 62\,400 \text{ lb}$$

SAMPLE PROBLEM 9.5
Temperature Change Problem

Problem

An aluminum block and a steel block fit snugly between two plates at 70°F (surfaces are in contact, but there is no stress). If the temperature of the blocks is raised to 270°F, what is the length of each block?

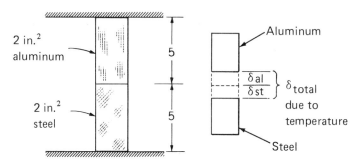

Solution

Step 1 If blocks are unrestrained:

$\delta_{al} = L \cdot \alpha \cdot \Delta_t = 5(13 \times 10^{-6})(200) = 0.0130$ in.
$\delta_{st} = L \cdot \alpha \cdot \Delta_t = 5(6.5 \times 10^{-6})(200) = 0.0065$ in.
$\delta_{total} = 0.0195$ in.

Step 2 We now consider the blocks compressed back to 10 in. *Note:* δ_{al} due to compression will *not* equal δ_{al} due to temperature change. The same applies to steel.

$\delta_{total} = \delta_{al} + \delta_{st}$
$F_{al} = F_{st}$

$$0.0195 = \frac{F_{al}L_{al}}{A_{al}E_{al}} + \frac{F_{st}L_{st}}{A_{st}E_{st}} = \frac{F(5)}{2 \times 10(10)^6} + \frac{F(5)}{2 \times 30(10)^6}$$

$$F = 19.5(10)^{-3}(10)^6 \times \frac{12}{4} = 58\,500 \text{ lb}$$

$$\delta_{al} = \frac{FL}{AE} = \frac{58\,500(5)}{2 \times 10(10)^6} = 0.0146 \text{ in.}$$

$$\delta_{st} = \frac{FL}{AE} = \frac{58\,500(5)}{2 \times 30(10)^6} = 0.0049 \text{ in.}$$

Step 3 To obtain the final lengths, the above deformations must be subtracted from the unrestrained length at 270°F:

Aluminum: $L_f = 5.0130 - 0.0146 = 4.9984$ in.

Steel: $L_f = 5.0065 - 0.0049 = 5.0016$ in.

SAMPLE PROBLEM 9.6
Temperature Change Problem

Problem

We will consider the setup in Sample Problem 9.3. If the nut is run up hand-tight at 70°F, what will the force on the bolt be (a) at 250°F and (b) at 0°F?

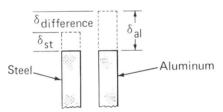

FBD (deformations due to temperature change)

Solution (a)

Step 1 If the members are unrestrained:

$\delta_{st} = L \cdot \alpha \cdot (T_f - T_i) = 9(6.5 \times 10^{-6})(180°) = 0.0106$ in.

$\delta_{al} = L \cdot \alpha \cdot (T_f - T_i) = 5(13 \times 10^{-6})(180°) = 0.0117$ in.

From the FBD, we can see that any stress is due to the difference in temperature strains.

$\delta_{diff} = 0.0011$ in. $= \delta_{total}$

We now have a problem similar to Sample Problem 9.3.

$$\delta_{total} = \frac{F_{st}L_{st}}{A_{st}E_{st}} + \frac{F_{al}L_{al}}{A_{al}E_{al}}$$

$$F_{al} = F_{st}$$

$$0.0011 = \frac{F(9)}{2.3(30 \times 10^6)} + \frac{F(5)}{6(10 \times 10^6)}$$

$$0.0011(10)^6 = 0.1305F + 0.0834F$$

force on bolt = F = 5140 lb

Solution (b)

There would be no force on the steel bolt at 0°F because the aluminum changes length twice as fast as steel and the bolt is not twice as long as the aluminum prisms.

METRIC SAMPLE PROBLEM 9.7
Length and Temperature Change

Problem

A bar of steel 2.0000 m long is heated 120°C above its normal temperature. What is the length at the higher temperature?

Solution

From the Appendix, the thermal coefficient of expansion is (12×10^{-6}) m/m/°C. Therefore:

$$\delta = \alpha \cdot \Delta_t \cdot L = (12 \times 10^{-6})(120)(2) = 2880(10)^{-6} \text{ m} = 2.88(10)^{-3} \text{ m}$$

new length = 2.0029 m

SUMMARY

When statics equations are not sufficient to solve for forces and reactions, stress–strain relationships may help. A useful formula is:

$$\delta = \frac{F \cdot L}{A \cdot E}$$

The deformation relationship of two materials in problems described in this chapter can be placed in one of the following categories:

$\delta_a = \delta_b$

$\delta_a = \delta_b \pm \text{constant}$

$\delta_{total} = \delta_a \pm \delta_b$

The formula for deformation due to temperature change is:

$\delta = \alpha \cdot \Delta_t \cdot L$

The temperature problems in this chapter can be solved in two parts. First, determine the unrestrained deformation due to temperature change. Then use the stress–strain relationships as outlined in the first section of this chapter.

PROBLEMS

Statically Indeterminate Machine Members with Axial Loads

1. Using the formulas $\sigma = F/A$, $\delta = \epsilon \cdot L$, and $E = \sigma/\epsilon$, derive the formula: $\delta = F \cdot L/A \cdot E$.
2. Refer back to Figure 9.1. The blocks are 20 in. high, and each is 3 in.² in area. The total load is 30 000 lb. Block A is steel, and block B is brass. Find the load on each block.
3. Refer to Sample Problem 9.1. If the load is not known, but the total deformation is 0.008 in., find the total load.
4. A steel prism measuring 1.5 in. × 1.5 in. × 6 in. long supports an axial compressive load P. The deformation is 0.02 in.
   (a) Find load P.
   (b) If the prism were originally 3 in. long, what load would cause a deformation of 0.02 in.?
5. A copper block is placed on top of a steel block; together they support a 10 000 lb compressive load. Each block is 0.75 in.² in cross-sectional area and 4 in. long before loading.
   (a) What is the length of each prism under the load?
   (b) If the load were not known, but the total length was 7.999 in. under load, find the load.
6. Solve Sample Problem 9.2 if both wires are made of steel.
7. Refer to Figure 9.1. If block A is steel and block B is aluminum, what proportion of the total load will block A take?

178   CHAPTER NINE

8. Figure 9.2 shows a 2000 lb load supported by three wires, each 0.3 in.² in cross-sectional area. Assume the load remains horizontal.
   (a) If all the wires are the same length (30 ft with no load), what is the load on each wire and the length of each wire?
   (b) If only the copper wires are 30 ft long with no load, and all wires take an equal share of the load, how long should the steel wire be before loading?
9. Figure 9.3 shows three prisms and their no-load dimensions. Block A is brass and 1.75 in.² in area; blocks B are steel, and each is 1.5 in.² in area. The rigid load is 100 kips. What load does each block take?
10. Refer to Sample Problem 9.3. How far should the nut be turned down to give a stress of 30 000 psi in the steel bolt?
11. Figure 9.4 shows a cross-sectional view of a reinforced concrete pillar (22 in. outside diameter). The stress in the concrete is 2000 psi compressive.
    (a) What is the stress in the steel reinforcing rods?
    (b) What strain results in each material?
    (c) What is the total load on the pillar?
    Use $4.5(10)^6$ psi for the modulus of elasticity of concrete.
12. (SI Units) In a setup similar to that in Figure 9.1, a hollow aluminum cylinder (300 mm outside diameter and 100 mm in diameter) and a solid brass cylinder 100 mm in diameter have an unloaded length of 800 mm. Determine the deformation and the load on each cylinder when they support a 10 Mg load.

Machine Members Subjected to Temperature Change

Note: Refer to the Appendix for the coefficients of expansion.

13. Figure 9.5 shows a bar 12 in. long at 70°F. A heating coil is wrapped around it. Assume the bar can expand only to the right. At what temperature will the light go on (a) if the bar is steel, and (b) if the bar is aluminum?
14. How much longer is a 1500 ft steel mine hoist cable in the summer at 90°F than in the winter at 30°F?

Statically Indeterminate Machine and Structural Members   179

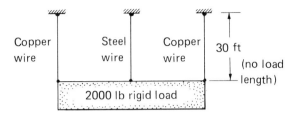

FIGURE 9.2   Problem 8

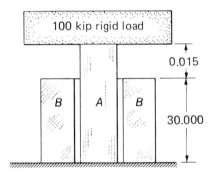

FIGURE 9.3   Problem 9

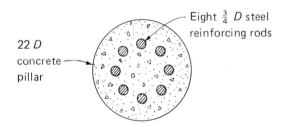

FIGURE 9.4   Problem 11

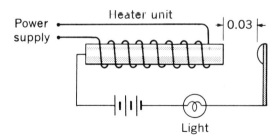

FIGURE 9.5   Problem 13

15. Refer to Figure 9.6. Both prisms are 0.83 in.² in cross section. The dimensions given are for 70°F. Both prisms are steel.
   (a) At what temperature will they make contact?
   (b) At what temperature will the stress be 10 000 psi?
   (c) Repeat (a) and (b). Assume the right-hand prism is copper.

16. A 1/16 in. diameter solid steel wire 3.0 ft long at 70°F is tied to a 50 lb weight. *Condition (a):* The wire supports the 50 lb as the weight touches, but does not exert a force on, the floor. *Condition (b):* The weight is resting its full 50 lb on the floor, but there is no slack in the wire. For each of the conditions, what should the temperature of the wire be to raise the weight 0.01 in. from the floor?

17. Refer to Sample Problem 9.5, which shows two prisms just touching at 70°F. Assume the temperature is raised to 160°F.
   (a) What force is exerted on each prism?
   (b) How long will the steel block be?
   (c) How long will the aluminum block be?
   (d) Sketch the position of the interface in relation to its original position.

18. Refer to Figure 9.7. The unloaded length of the wires must not change more than 0.2 in. when the load is added and there is a 40°F temperature increase. Assume the load remains level. How long can the wires be?

19. In Figure 9.8, block A is a 1 in.² prism of aluminum and block B is a 1.5 in.² prism of bronze. The pointer rod is connected as shown and is horizontal at 70°F. Where will the pointer be with a 50° increase in temperature?

20. Figure 9.9 shows a steel chain link designed for a shrink fit over the posts. Assuming the posts are rigid, how much above 70°F must the link be heated to fit over the posts?

21. The dimensions given in Figure 9.10 are for 70°F.
   (a) To what temperature must the thin brass ring be heated to slip over the post?
   (b) If the ring has a cross-sectional area of 0.04 in.², what tensile stress is on the cross section? Assume the post maintains its 5 in. diameter.
   Hint: You may consider the ring circumference equal to L.

22. (Extra Credit) A steel bolt, threaded its entire length and having a minor thread diameter of 0.4 in., is run through a copper tube (cross-sectional area 0.361 in.²) 6 in. long. A nut is turned down until it just touches the tube. What will be the stress on the bolt cross section if the temperature is raised 100°F?

Statically Indeterminate Machine and Structural Members   181

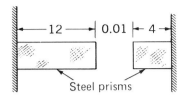

FIGURE 9.6   Problem 15

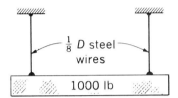

FIGURE 9.7   Problem 18

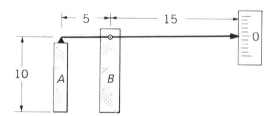

FIGURE 9.8   Problem 19

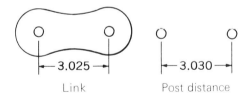

FIGURE 9.9   Problem 20

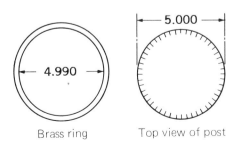

FIGURE 9.10   Problem 21

23. (Extra Credit, SI Units) A copper rod 500 mm long and 75 mm in diameter is wedged between two supporting walls. If there is no stress at the present temperature, what is the stress in the rod when its temperature is raised 30°C?

CASE STUDY

In the early twenties, the roof of a theater collapsed. One area of investigation was the bearing supports for the roof trusses. The original design called for brick piers measuring 28 in. × 16 in., as shown in Figure 9.11(a). Each steel truss rested on a steel pad placed on top of the pier. The pad was 20 in. × 16 in. and was placed as shown in Figure 9.11(a).

On investigation, it was found that the contractor had constructed a cavity in the top of each pier and filled it with concrete. The cavity was 20 in. long, 8 in. wide, and 20 in. deep. See Figure 9.11(b).

Use the following information to determine (a) if the original design met specifications and (b) if the actual construction met specifications:

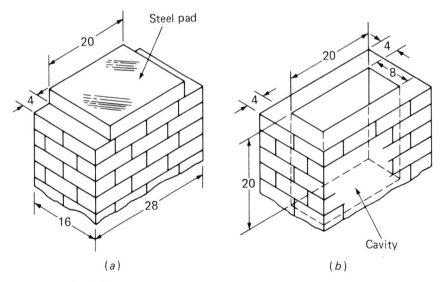

FIGURE 9.11 Bearing Supports for Theater Roof Trusses

load on each pier = 80 000 lb, and was evenly distributed across steel pad area

allowable stress for bearing on brick wall = 250 psi

allowable stress for bearing on concrete = 875 psi

modulus of elasticity for brick and mortar = 2 500 000 psi

modulus of elasticity for concrete = 4 500 000 psi

COMPUTER PROGRAM

The following program will solve problems similar to Sample Problem 9.1. The lengths of each part and the deformations must be equal. For example, refer to Sample Problem 9.1 and suppose the 1000 kip load is applied to a similar steel and concrete structure. However, assume that the steel has a maximum allowable compressive stress of 22 000 psi and the concrete has an allowable compressive stress of 3000 psi. Find the outside diameter required by increasing it in increments of 1/2 in. You may also want to change the steel thickness in increments of 1/4 in. Try using this computer program to check the answers to applicable end-of-chapter problems.

```
10   REM  AXIALLOADS
20   PRINT "STATICALLY INDETERMINATE AXIAL LOADS. THIS PROGRAM WILL CALCUL
     ATE"
21   PRINT "THE SHARE OF A LOAD SUPPORTED BY EACH OF TWO MEMBERS.  THE LEN
     GTHS"
22   PRINT "AND DEFORMATIONS OF EACH MEMBER TO EQUAL THE OTHER - AS IN "
23   PRINT "SAMPLE PROBLEM 9.1.  TYPE DATA IN THIS ORDER: TOTAL LOAD (LB)"
24   PRINT "LENGTH OF PART (IN.), 'E' FOR PART #1, 'E' FOR PART #2,"
25   PRINT "AREA FOR PART #1, AREA FOR PART #2."
30   INPUT P,S,E1,E2,A1,A2
40   LET K = (A1 * E1) / (A2 * E2)
50   LET F1 = P * K / (1 + K)
60   LET F2 = P - F1
70   LET D = F1 / (A1 * E1)
80   PRINT
85   PRINT
90   PRINT "FORCE ON PART #1 = "; INT (F1);" LB."
95   PRINT
100  PRINT "FORCE ON PART #2 = "; INT (F2);" LB."
105  PRINT
110  PRINT "DEFORMATION = ";D;" IN."
```

10

Thin-Walled Pressure Vessels

OBJECTIVES

This chapter demonstrates the theory behind the two principal formulas for determining the tensile stresses in thin-walled pressure vessels. The chapter's sample problems show you how to substitute the proper values in the formulas and apply these formulas correctly.

INTRODUCTION

In the early days of steam, boiler pressures were on the order of 60 psi, about the pressure in a bicycle tire. Many large modern boilers are designed to operate at pressures well over 3000 psi. In order to understand the relationship between internal pressure and the stresses in the wall of the vessel, we must first look at a bolted cover.

BOLTED COVERS OF PRESSURE VESSELS

Consider the problem of determining the size of bolts needed to hold the lid on the pressure vessel shown in Sample Problem 10.1. The maximum allowable tensile stress (σ_{all}) for the bolt material is 20 000 psi. Three bolts are to be used. The internal gage pressure (p) in the vessel is 300 psi. Note that both stress and internal pressure are measured in pounds per square inch, but are equal neither to each other nor to equivalent values that can be substituted for each other. The internal diameter of the vessel is 18 in. You will recognize that the total force exerted on the lid by the internal pressure is obtained by multiplying the pressure by the area of the lid. Therefore, Step 1 in the sample problem is

$$F_{total} = p \cdot A_{lid} = \frac{p\pi D^2}{4}$$

The FBD of the lid is shown in the sample problem, indicating that the force F_{total} is opposed by the three forces applied to the lid by the bolts. Also, the force per bolt must equal the allowable stress times the minor area of the bolt. These mathematical relationships are written in Step 2. By rearranging the equation in Step 3, we obtain the formula for the bolt area in Step 4. The area of maximum tensile stress is at the minor diameter of the bolt. Referring to the table of standard bolt sizes in the Appendix, we see that three 1½ in. bolts are required.

SAMPLE PROBLEM 10.1
Bolts for a Lid

Problem

Given the following data, find the size of bolts needed. Note that in the sketch, three bolts hold the lid in place.

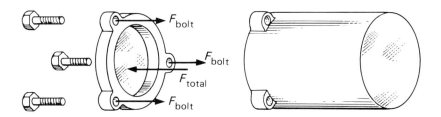

Data

F_{total} = total force on lid due to internal pressure
F_{bolt} = force on bolt
p = internal pressure = 300 psi
D = internal diameter = 18 in.
σ_{all} = allowable stress in bolt material = 20 000 psi
A_{bolt} = minor area of bolt

Solution

Step 1 The total force on the lid is equal to the pressure times the area of the lid:

$$F_{total} = \frac{p\pi D^2}{4}$$

Step 2 The total force is resisted by the three bolts:

$$F_{total} = 3F_{bolt} = 3(\sigma_{all} \cdot A_{bolt})$$

Step 3 Set the formulas in Steps 1 and 2 equal to each other:

$$3(\sigma_{all} \cdot A_{bolt}) = \frac{p\pi D^2}{4}$$

Step 4 Rearrange the preceding equation to solve for bolt area:

$$A_{bolt} = \frac{p\pi D^2}{4 \cdot 3 \cdot \sigma_{all}} = \frac{300 \cdot 3.147 \cdot (18)^2}{4 \cdot 3 \cdot 20\,000} = 1.275 \text{ in.}^2$$

Answer

Step 5 From the table of bolt sizes in the Appendix, choose the 1½ in. bolts.

TENSILE STRESS ON A CIRCUMFERENTIAL SEAM OR JOINT

Let us now remove the bolt assembly in the previous illustration and weld the lid to the cylinder wall, the weld being of the same thickness as the cylinder wall. The problem will be to determine the tensile stress in the weld.

Procedure for Deriving the Stress Formula for a Circumferential Seam

Step 1 The sum of the resisting forces in the circumferential seam equals the force on the lid due to internal pressure.

$$\Sigma F = F_{total}$$

Step 2 The resisting forces are related to the tensile stress in the seam.

$$\sigma \cdot A_{seam} = p \cdot A_{lid}$$

Step 3 Substitute components for areas.

$$\sigma(\pi D t) = \frac{p \pi D^2}{4}$$

Step 4 Rearrange the preceding equation.

$$\sigma = \frac{pD}{4t}$$

The FBD of the lid is drawn in Figure 10.1(a). The area of the weld is the area of the annular ring described by the cylinder wall.

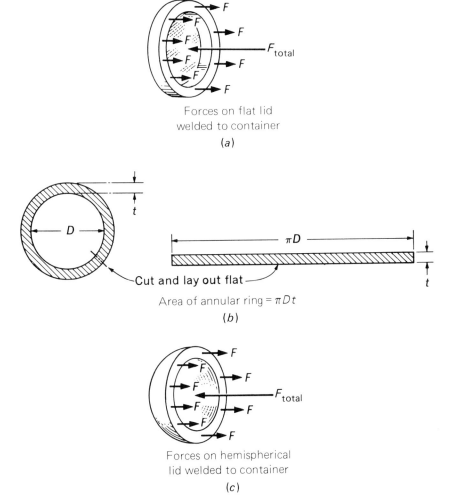

Forces on flat lid
welded to container
(a)

Area of annular ring = $\pi D t$
(b)

Forces on hemispherical
lid welded to container
(c)

FIGURE 10.1 Circumferential Weld Area on a Lid

This area is πDt, as shown in Figure 10.1(b) and in Step 3 of the procedure. D is the inside diameter of the cylinder, t is the wall thickness, and A_{seam} is the area of the weld or any circumferential seam in a cylinder. This formula for the area of an annular ring is one of two approximations that can be made with negligible error, because a thin-walled cylinder is one whose wall thickness is less than 1/10 of the inside diameter.

By rearranging the equation in Step 3, we obtain Step 4, which is the formula for the tensile stress on a circumferential seam (joint) in a thin-walled vessel:

$$\sigma = \frac{pD}{4t}$$

The second approximation results from the assumption that the stress does not vary but is constant across the thickness of the wall, and the general formula

$$\sigma = \frac{F}{A}$$

can be used. Figure 10.2 shows the true stress distribution across the wall of a cylinder with an internal pressure. However, for thin-walled vessels, the maximum stress is very close to the average stress value.

An interesting question now arises: If the lid is hemispherical in shape, as in Figure 10.1(c), instead of flat, will the forces and our calculations remain the same? The answer is yes. In fact, the lid could have curvatures other than spherical. To demonstrate this, let us look at the small area A' near the top of the soap bubble in Figure 10.3(a). The area is small enough to be considered flat and is at some angle, α, to the horizontal. The internal pressure p is considered to be per-

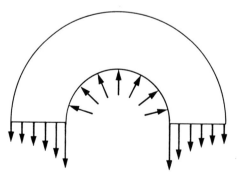

FIGURE 10.2 Stress Distribution in Pressure Vessel Wall

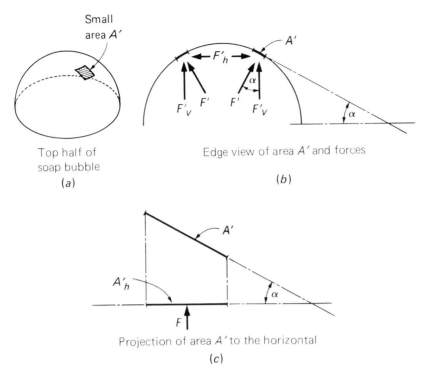

FIGURE 10.3 Force Exerted on a Hemispherical Surface

pendicular to the bubble's surface at all points. This means that the force F' on this small area is perpendicular to the area and is equal to $p \cdot A'$. The force F' is now replaced by its horizontal and vertical components. The horizontal component is not considered because it will be cancelled by a like horizontal force on some area placed symmetrically on the other side of the bubble, as shown in Figure 10.3(b). Note that the horizontal components (F'_h) of symmetrically placed areas cancel each other, but because the vertical components (F'_v) are additive, they must be considered. The magnitude of the vertical component is obtained by the following relationship:

$$F' = p \cdot A'$$
$$F'_v = F' \cdot \cos \alpha = p \cdot A' \cdot \cos \alpha$$

Keeping this relationship in mind for later reference, we now proceed on an alternative approach. Figure 10.3(c) illustrates the projection of the area A' onto a horizontal surface. The horizontal projected area A'_h is given by the formula:

$$A'_h = A' \cdot \cos \alpha$$

The force F on the area A_h is given by the formula:

$$F = p \cdot A'_h = p \cdot A' \cdot \cos \alpha$$

This is the same formula that was obtained for the force F'_v previously. Therefore, since $F = F'_v$, we may use the *projected plane area* of the curved surface in the formula:

$$F = F_{total} = p \cdot A$$

where A is the projected plane area. The total force acting in a vertical direction on the top half of the soap bubble $= \Sigma F = \Sigma p \cdot A'_h$, or $F_{total} = p \cdot A$, where A is the projected plane area of the bubble.

Thus, the formula for a circumferential seam, $\sigma = pD/4t$, is to be used for determining the tensile stress on any circumferential joint, including the circular joint of a spherical vessel where the center of the circular joint is at the center of the sphere, called a *great circle*.

TENSILE STRESS ON A LONGITUDINAL SEAM OR JOINT

A more critical tensile stress appears on longitudinal seams of cylindrical vessels. In order to understand this stress, we must visualize that the pressure in the vessel is tending to separate the vessel into two halves, as shown in Figure 10.4(a). The FBD for half of the cylindrical tank is shown in Fig. 10.4(b). The statics equilibrium formula shows us that

$$F + F = F_{total}$$

Also, the force on each wall is equal to the stress times the wall area:

$$F = \sigma \cdot A_{wall} = \sigma \cdot L \cdot t$$

where
$t = $ wall thickness
$L = $ length of cylinder in inches

As in the case with the circumferential seam, the force F_{total} is

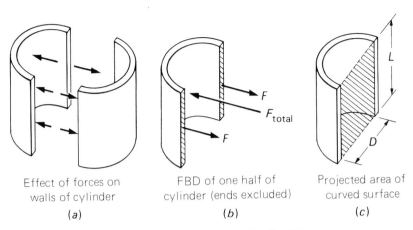

FIGURE 10.4 Forces on a Longitudinal Seam

equal to the pressure times the projected plane area of the curved surface; see Figure 10.4(c):

$$F_{total} = p \cdot A = p \cdot L \cdot D$$

where
- A = projected plane area of curved surface in square inches
- L = length of cylinder in inches
- D = diameter of cylindrical vessel in inches

Substituting in the equilibrium formula ($F + F = F_{total}$), we obtain:

$$\sigma \cdot L \cdot t + \sigma \cdot L \cdot t = p \cdot L \cdot D$$
$$2(\sigma \cdot L \cdot t) = p \cdot L \cdot D$$

Rearranging, we have:

$$\sigma = \frac{pD}{2t}$$

This is the formula for the tensile stress in the longitudinal cross section of the wall (or the longitudinal seam).

As mentioned earlier, the tensile stress in a cylinder wall is considered to be constant across the thickness of the wall, when in actual fact it is not. However, if the thickness of the wall is less than 1/10 of the inside diameter, the maximum stress would be about 10 percent greater than the average stress. This percentage decreases as the wall thickness becomes a smaller percentage of the vessel diameter. The formula above is considered sufficiently accurate that the more complicated solution for thick-walled cylinders need not be used. Metric

192 CHAPTER TEN

PHOTO 10.1 Natural Gas Storage Tanks of the Boston Gas Company

Sample Problem 10.2 uses the formula for stress on a longitudinal seam.

METRIC SAMPLE PROBLEM 10.2
Stress on a Longitudinal Seam

Problem

Determine the stress on the longitudinal seam of a pressure vessel 3.5 m long, 2.4 m in diameter, and with a wall thickness of 10 mm. The pressure is 560 kPa.

Solution

Note: The length of the vessel makes no difference. Diameter and thickness must be in metres.

wall thickness = 10 mm = 0.010 m

$p = 560(10)^3$ Pa

$$\sigma = \frac{pD}{2t} = \frac{560(10)^3 \,(2.4)}{2(0.010)} = 67.2(10)^6 \text{ Pa} = 67 \text{ MPa (rounded)}$$

SUMMARY

The two principal formulas for stresses in thin-walled pressure vessels are as follows:

1. For a circumferential cross section of the cylinder wall and for great circle wall cross sections on spherical vessels,

$$\sigma = \frac{pD}{4t}$$

2. For longitudinal wall cross sections of cylindrical vessels,

$$\sigma = \frac{pD}{2t}$$

PROBLEMS

Bolted Covers and Circumferential Seams

1. (a) Solve Sample Problem 10.1, but change the internal pressure to 100 psi.
   (b) Given the information in Sample Problem 10.1, determine the number of 3/4 inch bolts required.
2. The lid in Sample Problem 10.1 is welded to the container. The wall thickness is 1/2 in. Determine the stress in the weld (seam).

3. In Figure 10.5, the inspection cover is secured in place with ten 1¼ in. bolts. The allowable tensile stress in the bolts cannot exceed 15 000 psi. What is the allowable pressure in the vessel?
4. A spherical gas tank 50 ft in diameter has a wall thickness of 1/2 in. If the allowable stress in the wall is 20 000 psi, what pressure can the tank hold?
5. An expansion chamber is welded to a pipeline carrying 350 psi of pressure. Using the dimensions in Figure 10.6, determine the tensile stress in the weld.

Longitudinal Seams

6. A home pressure cooker is designed to operate at 15 psi pressure. The inside diameter of the cooker is 9 in.
   (a) What is the force on the cover?
   (b) If the pressure regulator fits over an opening 0.10 in. in diameter, what must the regulator weigh to control the pressure?
   (c) Using an allowable tensile stress of 20 000 psi, determine the wall thickness.
7. The maximum allowable tensile stress for copper tubes in service not exceeding 350°F is 4500 psi. What wall thickness is required for a tube with an inside diameter of 4 in. and with an internal pressure of 200 psi?
8. A 36 in. Class E cast iron pipe, strong enough to withstand a 500 ft head of water, is required to have a wall thickness of 1.80 in. What is the tensile stress on a longitudinal seam?
9. We are required to design the water tank illustrated in Figure 10.7. The allowable tensile stress in the steel plates is 18 000 psi. What thickness must the bottom row of plates be?
10. (Extra Credit) Figure 10.8 shows a wooden vessel designed to hold a fluid with a specific gravity of 1.80. The wooden walls are contained by steel bands 1 in.² in cross-sectional area, and the bands are placed 1 ft apart. The lowest band is placed 1/2 ft above the bottom of the container. Determine the stress in the lowest steel band.
11. Determine the change (if any) in stress in the longitudinal and circumferential seams if:
    (a) A given cylinder has its length doubled and the pressure remains the same.
    (b) A given cylinder has its diameter doubled while the pressure remains the same.
    (c) A given cylinder has its length and pressure doubled.

Thin-Walled Pressure Vessels   195

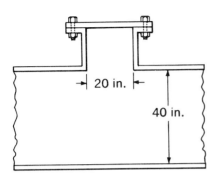

FIGURE 10.5   Problem 3

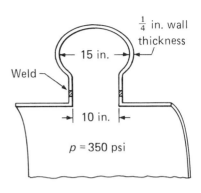

FIGURE 10.6   Problem 5

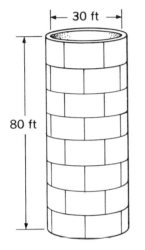

FIGURE 10.7   Problems 9 and 12

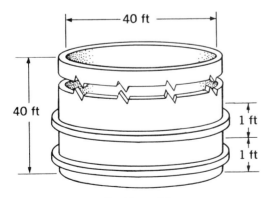

FIGURE 10.8   Problem 10

(d) The diameter of a spherical tank is doubled while the pressure remains the same.

12. (SI Units)   A natural gas storage tank similar to that in Figure 10.7 is 30 m high and 25 m in diameter. The wall thickness is 21 mm. If the allowable stress in the wall is 110 MPa, what is the allowable tank pressure?

13. (SI Units)   A spherical gas tank 15 m in diameter has a wall thickness of 15 mm. If the allowable stress in the wall is 80 MPa, what pressure can the tank take?

14. (SI Units)   If the stress calculated in the metric Sample Problem 10.2 cannot be exceeded, and the pressure is raised to 10 MPa, calculate the required wall thickness.

COMPUTER PROGRAM

The following program is useful for problems involving vertical storage tanks. For example, using the data given in Problem 9 and the fact that each of the welded steel plates is 4 ft high, calculate the thickness required for each row of plates. You may want to adjust the computer program so that you do not have to keep typing in repetitive data.

```
10   REM   PRESSURE
20   PRINT "THIS PROGRAM IS FOR A CYLINDERICAL STORAGE TANK, PLACED VERTIC
     ALLY"
21   PRINT "AS IN FIGURES 10.7 AND 10.8."
22   PRINT
23   PRINT "THE 'LONGITUDINAL SEAM' FORMULA AND THE 'PRESSURE DUE TO THE H
     EIGHT"
24   PRINT "OF A LIQUID' FORMULA ARE USED.
25   PRINT
30   PRINT "TO FIND PRESSURE-TYPE 1, TO FIND STRESS-TYPE 2, TO FIND THICKN
     ESS-TYPE 3."
40   INPUT A
50   IF A = 1 THEN   GOTO 100
60   IF A = 2 THEN   GOTO 200
70   IF A = 3 THEN   GOTO 300
80   END
100  HOME
110  PRINT "TYPE, IN ORDER:HEIGHT OF TANK(FT) AND SPECIFIC GRAVITY OF LIQ
     UID."
111  PRINT "
120  INPUT H,SG
130  LET P =   INT (H * .43333 * SG * 100 + .51) / 100
140  PRINT
150  PRINT "PRESSURE AT A DEPTH OF ";H;" FT = ";P;" PSI."
160  GOTO 80
200  HOME
210  PRINT "TYPE, IN ORDER: HEIGHT OF TANK (FT), SPECIFIC GRAVITY OF LIQU
     ID,"
```

```
211  PRINT "DIAMETER OF TANK (FT), AND WALL THICKNESS (IN.)."
220  INPUT H,SG,D,T
230  LET P = H * .43333 * SG
240  LET S = (P * D * 12) / (2 * T)
250  PRINT
260  PRINT "STRESS AT A DEPTH OF ";H;" FT = "; INT (S * 10 + .51) / 10;"
     PSI."
270  PRINT
280  GOTO 150
300  HOME
310  PRINT "TYPE, IN ORDER: HEIGHT OF TANK (FT), SPECIFIC GRAVITY OF LIQU
     ID,"
311  PRINT "DIAMETER OF TANK (FT), AND ALLOWABLE STRESS (PSI)."
320  INPUT H,SG,D,S
330  LET P = H * 0.43333 * SG
340  LET T = P * D * 12 / (2 * S)
350  PRINT
360  PRINT "WALL THICKNESS AT DEPTH ";H;" = "; INT (T * 1000 + .51) / 100
     0;" IN.
370  PRINT
380  GOTO 150
```

11

Welded and Riveted Joints

OBJECTIVES

When you have learned the steps and procedures involved in solving welded and riveted joint problems, you will be able to use your knowledge of statics and strength of materials to analyze forces on joints and to specify the required length of fillet welds.

INTRODUCTION

For centuries, blacksmiths have forge-welded metals by heating and pounding them together. But the welding techniques used by industry today have a very recent history. In 1885, the first patent was granted for arc welding, and gas welding was developed shortly after. In the early days, welding was used for repairing; the test of a good weld was to drop the repaired part on the floor. Since the 1930s, however, welding has developed rapidly as an important engineering tool in the manufacturing and construction industries.

The riveting of metals is probably as old as blacksmithing. Museums display ancient armor used by knights. When viewing the armor, you will notice that the pieces of armor are riveted together. What has changed is our knowledge of the forces involved and the tools and materials used.

WELDED JOINTS

Metal plates are generally joined together by *butt joints* or *lap joints* (see Figure 11.1). Butt joints are formed when the plates are butted

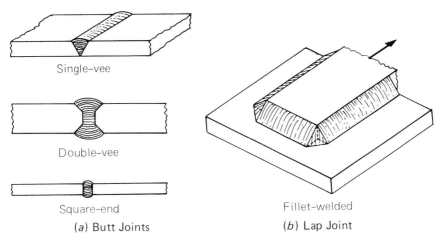

FIGURE 11.1 Types of Welds

together and welded at the joint, as illustrated in Figure 11.1(a). In general, square-end butt joints are used for the thinnest plates; single-vee, for thicker plates; and double-vee, for the thickest plates. Also, other things being equal, a double-vee butt joint is generally considered to be stronger than the other types.

The weld metal may be built up above the level of the plates for reinforcement. However, there are limits to the buildup; too much will create concentrated stresses in the adjacent metal and tend to reduce the strength of the joint. The accepted practice in determining the load a butt weld will take is to consider only the cross-sectional area of the thinnest plate being joined. The allowable stress that may be placed on a butt weld will vary from 100 percent of the strength of the plate to 50 to 60 percent.

The efficiency of the butt-welded joint is set forth in specifications determined by the American Welding Society and generally applied to shop work, such as boiler fabrication. Factors considered when rating the efficiency of a butt-welded joint are as follows:

1. Type of loading, such as tensile, compressive, shear, bending, steady, varying, shock, and combinations of these.
2. Type of inspection or thoroughness of inspection to determine the quality of weld. In critical situations, the weld may be X-rayed.
3. Whether or not the butt joint is square, vee, or double-vee.
4. Type of material in the main plates and in the weld metal.
5. Whether or not stress-relieving is applied to the weld. High heat and rapid cooling of the welding process set up localized

stresses in the adjacent plate material. Generally, this is unimportant in low-carbon steels, but high-carbon steels must be stress-relieved if the joint is to perform satisfactorily.

For solving problems, you may apply the efficiency to the stress or the load. For instance, given a 60 percent efficient joint with an allowable stress of 13 600 psi, you may use a stress value of 13 600 × 0.6 = 8200 psi. Or, you may calculate the load the joint can take using the allowable stress, and then multiply the answer by 0.6.

For the fabrication of steel in the field on construction projects, the allowable tensile, compressive, or shear stress limits on welds are the same as for the base metal of the structural members. There are some exceptions, but for solving problems in this chapter, you should use the allowable stress values for metals given in the Appendix.

Lap joints are formed by lapping one metal plate over another and joining them with welds placed along the sides of one plate and/or across the ends of the plates, as illustrated in Figure 11.1(b). This type of weld is called a *fillet weld*. If we look at a free body diagram of a fillet-welded lap joint (see Figure 11.2), it can be seen that the weld surfaces in contact with the plates are subjected to shear stresses. As brought out in previous chapters, for a given force, the maximum shear stress occurs on the smallest area parallel to the applied force. Since the weld cross section is considered to approximate the shape of a 45° right triangle, it will be noted that the smallest area (the critical area) is determined by the length of the weld multiplied by the throat dimension of the weld. The position of the throat dimension is shown in Figure 11.2(d). End fillet welds are handled in the same manner as side fillet welds, although it can be shown both mathematically and experimentally that the end weld is slightly stronger. For fillet welds that are convex in shape, the throat dimension is measured only to the line drawn between the ends of the legs. Any extra strength due to the convex shape is not considered. On the other hand, if the weld should be concave in shape (not generally recommended), the throat dimension is measured to the surface of the weld. A concave shape usually means a weak weld in proportion to the amount of weld metal used.

In order to determine the length of welds and their positions or placement, it becomes necessary to reemphasize an assumption made for almost all problems:

> *Unless otherwise stated, the load, or the resultant of a distributed load, is assumed to be applied along the centroidal axis of the member.*

Therefore, if the centroidal axis is equidistant from the edges of the member, as illustrated in Figure 11.2, the welds along each edge of

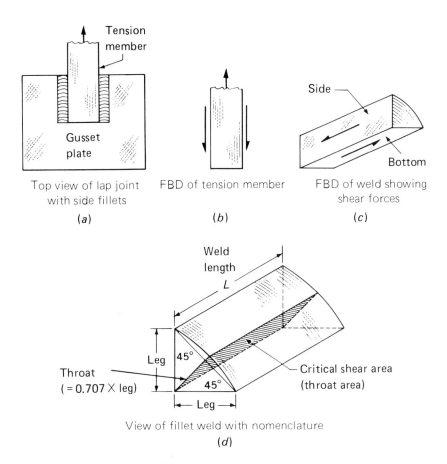

FIGURE 11.2 Fillet Weld

the plate would be the same length. However, if the centroidal axis is not symmetrical with respect to the edges, as is the case for the angle member in Sample Problem 11.1, then the proportioning of the weld lengths along each edge must be solved by the summation of moments.

The allowable shear stress for a fillet weld is specified by the American Welding Society and varies depending on the welding rod metal used and whether the structure is a bridge or a building. Sample Problem 11.1 uses a value of 13 600 psi for the allowable shear stress. This was a common value a number of years ago. For the problems at the end of this chapter, use the allowable shear stress for the metals being used in the connection. The stress values may be found in the Appendix.

Procedure for Solving Welded Joint Problems

Step 1 Draw a free body diagram of one plate, indicating the force the weld must resist.

Step 2 For fillet welds, determine the throat area of the weld. For butt welds, use the cross-sectional area of the plate and determine the required weld length.

Step 3 If the loading is eccentric, solve for the force on each weld by moment summation ($\Sigma M = 0$).

SAMPLE PROBLEM 11.1
Fillet-Welded Lap Joint

Problem

An L 4 × 3 × 1/2 angle member has its long side welded to a gusset plate. Use side fillets only with a 3/8 in. leg. The tensile load *P* is 57 000 lb. The allowable shear stress is 13 600 psi. Determine (a) the total length of the weld, (b) the length of weld *B* and weld *C*, and (c) how much overlap is needed.

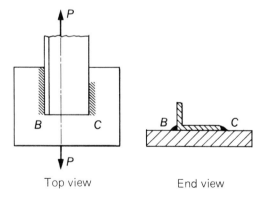

Top view   End view

Solution (a)

Step 1 First, an FBD is drawn of the angle iron. Handbook information gives the centroidal distance as 1.33 in. from the angle edge. For sheer stress, use 13 600 psi.

Welded and Riveted Joints 203

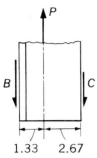

FBD of angle member

Step 2 Next, the throat area of the weld is determined (see shaded throat area in the sketch that follows) and then the force per inch of weld length. From this information, the total weld length is determined.

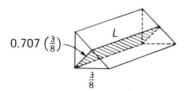

Sketch of fillet weld

$\Sigma F_y = 0 = P - (B + C)$

$B + C = 57\,000$ lb

$\tau = \dfrac{P}{A}$

$13\,600 = \dfrac{57\,000}{0.707(3/8) \times L}$

$L = 15.8$ in.

Note: The force per inch of weld length = 57 000/15.8 = 3610 lb/in.

Solution (b)

Step 3 Finally, a summation of moments is used to find the length of each weld. This is probably the most confusing step because the weld forces are shear forces and are not concentrated at some one point as in previous moment summations. You must mentally picture each weld applying a force to the member at one point. These points are in a line, as shown in the force diagram that follows. Once the forces are known, each weld length can be determined.

204   CHAPTER ELEVEN

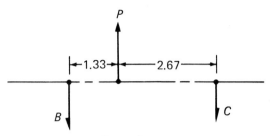

Force diagram

$\Sigma M_C = 0 = P(2.67 \text{ in.}) - B(4 \text{ in.})$

$B = 38\,000$ lb

length of weld $B = \dfrac{38\,000}{3610} = 10.55$ in.

$\Sigma M_B = 0 = P(1.33 \text{ in.}) - C(4 \text{ in.})$

$C = 18\,950$ lb

length of weld $C = \dfrac{18\,950}{3610} = 5.25$ in.

Solution (c)

overlap = 10.55 in.

RIVETED JOINTS

The metal members of structures are frequently connected by rivets or bolts. As in weld joints, the two most common types of connections are lap joints and butt joints. Figure 11.3 shows details of the connections. Note that the butt joint is made by putting the two main plates together and riveting one or two cover plates to transmit the load. The terms *single-*, *double-*, and *triple-riveted* refer to the number of rows of rivets in each main plate.

Figure 11.4 shows the critical areas stressed in a riveted joint. The cross section of each rivet is stressed in shear, as in Figure 11.4(*b*). Each half of a rivet has a bearing (compressive) stress applied, as shown in Figure 11.4(*c*). The critical bearing area is considered to be the plane area represented by $t \times d$. An equal bearing stress appears opposite the rivet in the plate. The net cross-sectional areas of plates are stressed

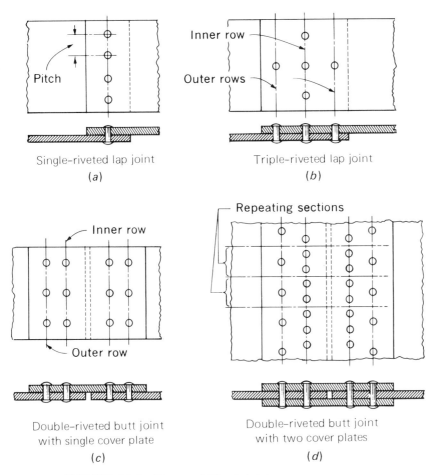

FIGURE 11.3 Types of Rivet Joints

in tension or compression, as portrayed in Figure 11.4(d). The shaded area in the sketch is the critical area, or net area, in tension. The formula for the net area, with two holes in the plate, is as follows: $A = (W - 2D)t$. Therefore, in order to analyze a riveted joint, the stresses and forces appearing on (1) the circular cross section of each rivet, (2) the projected plane area $t \times d$ of the rivet, and (3) the net cross section of the main plate must be compared. Allowable stresses for A36 structural steel are given in the Appendix.

The efficiency of a riveted joint is equal to the load the joint will take times 100 divided by the tensile (compressive) load the gross plate area will take. A riveted joint can never be 100 percent efficient

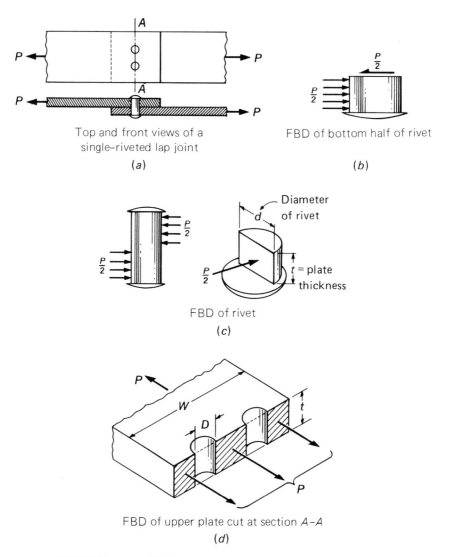

FIGURE 11.4 Critical Areas in Riveted Joints

because of the holes in the plates. The following assumptions and facts apply to riveted joint problems:

1. A basic assumption is that each rivet takes its fair share of the load as determined by the number of shear areas. Experiments have proven the soundness of this assumption.
2. To prevent the possibility of tear out and to simplify calcula-

tions, rivet holes should not be placed closer to the end of the plate than about two times the rivet diameter. *Tear out* occurs when the plate material ruptures in a direction toward the end of the plate instead of across the plate.
3. Boiler and pressure vessel fabrication is done in a shop where the rivet holes are drilled and matched and the rivets are set by machine. Because of these methods and close control, the rivet hole is taken as the same size as the rivet when making calculations.
4. Structural members may have their rivet holes punched out; this step weakens the material slightly around the hole. Also, conditions cannot be controlled as closely in the field as in the shop. For these reasons, it is standard practice to add 1/16 in. to the hole diameter to obtain the "effective" hole diameter when solving structural rivet problems. However, we will *not* do this for problems in this text. In order that you may concentrate on the basics, consider all hole or rivet diameters to be effective hole diameters.
5. There are concentrated stresses in the plate area around the rivet holes, as described in Chapter 8. However, specifications and codes frequently allow the technician or engineer to avoid this computation by lowering the allowable stress in this plate area. For the purpose of simplification, this text will use the same stress on plate cross sections with or without rivet holes.

Procedure for Solving Riveted Lap Joint Problems

Ideally, the three critical areas (net plate area, rivet shear area, and rivet bearing area) should all be stressed to their maximum safe values. But the use of standard sizes and materials means that these values probably will not be reached. Therefore, you must determine the safe load a joint can take when each critical area is stressed to its allowable value. Compare the loads you have calculated, and choose the lowest value as the maximum safe load for the joint.

Step 1 Refer to the free body diagrams in Figure 11.4 for each of the steps below where a critical area must be determined.

Step 2 Determine the number of shear areas on the rivets, and analyze for stress or force as specified.

Step 3 Determine the number of bearing areas on the rivets, and analyze.

CHAPTER ELEVEN

Step 4 Determine the net cross-sectional area of the smallest plate, and then solve for the stress or force as called for in the problem. If there is more than one row of rivets, the net cross-sectional area for each row may have to be analyzed.

Step 5 If the problem is to determine the load the joint can take, then the smallest of the loads calculated above is the answer.

Sample Problem 11.2 illustrates a riveted lap joint.

SAMPLE PROBLEM 11.2
Riveted Lap Joint

Problem

Given the following sketch and facts, determine (a) the allowable load P and (b) the efficiency of the joint.

$$\text{plates} = \frac{1}{2} \text{ in. thick}$$

$$\text{rivets} = \frac{3}{4} \text{ in. in diameter}$$

allowable σ_t = 20 000 psi

allowable σ_c = 32 000 psi

allowable τ = 15 000 psi

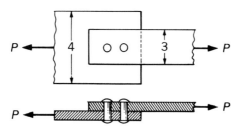

Solution (a)

Step 1 Refer to the free body diagram.

Step 2 The load as determined by shear areas is:

$$P = \tau A = 15\,000 \cdot \frac{\pi(3/4)^2 \cdot 2}{4} = 13\,250 \text{ lb}$$

Step 3 The load as determined by bearing areas is:

$$P = \sigma_c A = 32\,000 \cdot 2 \cdot \frac{3}{4} \cdot \frac{1}{2} = 24\,000 \text{ lb}$$

Step 4 The load as determined by tension area is:

$$P = \sigma_t A = 20\,000 \cdot \left(3 - \frac{3}{4}\right) \cdot \frac{1}{2} = 22\,500 \text{ lb}$$

Step 5 Thus, the allowable load is:

$$P = 13\,250 \text{ lb}$$

Solution (b)

The 3 in. plate with no holes can take a tensile load of:

$$\sigma_t A = 20\,000 \cdot 3 \cdot \frac{1}{2} = 30\,000 \text{ lb}$$

Thus, the efficiency of the joint is:

$$\text{efficiency} = \frac{13\,250 \cdot 100}{30\,000} = 44 \text{ percent}$$

Procedure for Solving Riveted Butt Joint Problems

The steps for solving riveted butt joint problems are essentially the same as for lap joints, with the added condition that the cover plates must be considered. If a free body diagram is drawn for each half of a butt joint, you can see that the loads on each half are identical and that half of the joint is resisting the other half. Therefore, only half of the joint need be considered. If the riveted joint is very long, as in a longitudinal boiler seam, then the strength of the joint is calculated for a repeating section rather than for the total length of the joint. Refer back to Figure 11.3(d). Note that the width of the repeating section is equal to the pitch of the outer row. See also Figure 11.3(a), and note that the pitch of a row of rivets is equal to the distance between the rivet center lines. These center lines were moved in Figure 11.3(d) to indicate clearly the number of rivets in each repeating section. Every repeating section must be identical to every other section so that the calculations for one apply to all.

210   CHAPTER ELEVEN

A look at Sample Problem 11.3 shows that only those forces on the left side of the joint need be considered because the right side is identical. The three rivets have a total of six shear areas. The cover plates have a total thickness of 1/2 in. Since the main plate has a 3/8 in. thickness, it will have a smaller area bearing against the rivets. Therefore, the main plate thickness is used for finding the bearing area load. The main plate net area at the inner row of rivets is less than the area at the outer row and, therefore, takes a smaller force—that is, 30 000 lb versus 37 500 lb. However, the inner row plate area never sees the full load P because it is assumed each rivet takes a fair share of the load, and the outer row rivet has taken off 1/3 of the load P. This means only 2/3 of P ever appears at the inner row (on the main plate). Therefore, as far as the inner row main plate area is concerned, P can equal 45 000 lb because only 30 000 lb will be applied to its area. The inner row of the cover plates must also be checked for tension.

SAMPLE PROBLEM 11.3
Double-Riveted Butt Joint

Problem

Given the following sketch and facts, determine (a) the allowable load P and (b) the efficiency of the joint.

rivets = 1 in. in diameter
allowable shear stress = 15 000 psi
allowable bearing stress = 32 000 psi
allowable tensile stress = 20 000 psi
main plates = $\frac{3}{8}$ in. thick
cover plates = $\frac{1}{4}$ in. thick

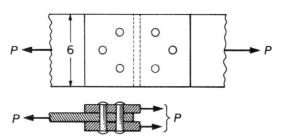

Solution (a)

Step 1 Refer to the free body diagram.

Step 2 *Note:* Rivets are in double shear. The joint load as determined by shear areas is:

$$P = \tau A = 15\,000 \cdot 6 \cdot \frac{\pi(1)^2}{4} = 70\,600 \text{ lb}$$

Step 3 *Note:* The bearing area of the main plates will be smaller than that of the cover plates. The joint load as determined by the bearing area is:

$$P = \sigma_c A = 32\,000 \cdot 3 \cdot 1 \cdot \frac{3}{8} = 36\,000 \text{ lb}$$

Step 4 The joint load as determined by the net area at the outer row is:

$$P = \sigma_t A = 20\,000 \cdot (6-1) \cdot \frac{3}{8} = 37\,500 \text{ lb}$$

The inner row plate tension load (not the joint load) is:

$$\sigma_t A = 20\,000 \cdot (6-2) \cdot \frac{3}{8} = 30\,000 \text{ lb}$$

This value is only 2/3 of the joint load because the rivet in the outer row takes 1/3 of the joint load. Therefore, the joint load is:

$$P = \frac{3}{2} \cdot 30\,000 = 45\,000 \text{ lb}$$

The joint load as determined by the cover plates in tension on the inner row is:

$$P = \sigma_t A = 20\,000 \cdot (6-2) \cdot \frac{1}{2} = 40\,000 \text{ lb}$$

Step 5 Choose the smallest value for the allowable joint load:

$$P = 36\,000 \text{ lb}$$

Solution (b)

For the main plate gross area in tension:

$$P = \sigma_t A = 20\,000 \cdot 6 \cdot \frac{3}{8} = 45\,000 \text{ lb}$$

Thus, the efficiency of the joint is:

$$\text{efficiency} = \frac{36\,000 \cdot 100}{45\,000} = 80 \text{ percent}$$

METRIC SAMPLE PROBLEM 11.4
Riveted Lap Joint

Problem

Two A36 steel plates, each 15 mm thick and 90 mm wide, are lap-joined with one row of two rivets. Rivet holes are 20 mm in diameter. Use the AISC code in the Appendix for allowable stress values. Determine the load the joint can take.

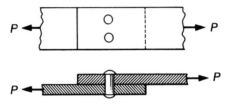

Solution

Step 1 The allowable stress values are as follows:

$\sigma_t = 150$ MPa

$\sigma_c = 600$ MPa

$\tau = 99$ MPa

Because stress values are in pascals (newton/metre^2), dimensions must be in metres and forces in newtons.

Step 2

$$P_t = \sigma_t A_t = 150(10)^6 [0.015(0.090 - 0.040)] = 113(10)^3 \text{ N}$$

Step 3

$$P_c = \sigma_c A_c = 600(10)^6 [2(0.020)(0.015)] = 360(10)^3 \text{ N}$$

Step 4

$$P_\tau = \tau A_\tau = 99(10)^6 \left[2\left(\frac{\pi(0.020)^2}{4}\right) \right] = 62(10)^3 \text{ N}$$

Step 5 The joint can take a load of 62 kN.

PROBLEMS

Welded Joints

Unless otherwise specified, use the allowable stresses from the Appendix for the metals being joined. When steel is called for, use data for A36 steel.

1. Two steel plates, 3/8 in. thick, are to be butt-welded, and the joint is expected to hold a 10 000 lb tensile load. How long should the butt weld be if it is (a) 100 percent efficient and (b) 80 percent efficient?

2. Two steel plates, 1/2 in. thick by 3 in. wide, are butt-welded. The joint is expected to take a tensile load.
   (a) What load can be applied if the weld is judged to be 85 percent efficient?
   (b) Consider this a segment of a longitudinal boiler seam, the boiler being 5 ft in diameter. What is the allowable pressure in the boiler? The joint is 85 percent efficient.

3. Construct a table to show the load a 1 in. long fillet weld can take for weld legs of 1/8 in., 1/4 in., 3/8 in., and 1/2 in. Consider the welded plates A36 steel.

4. Refer to Figure 11.1(b). If the top tension member of the assembly is 4 in. wide by 1/2 in. thick and the gusset plate is stronger:
   (a) What is the maximum allowable load this assembly can take? *Hint:* There is no limit to the amount of overlap.
   (b) What length of fillet weld is necessary?
   (c) If side fillets only are used, how much should the plates overlap?
   (d) How much should the plates overlap if side fillets and two end fillets are used?

5. Refer to Figure 11.2(a). A 3000 lb load is applied along an axis that is 1/2 in. from the left edge and 2 in. from the right edge of the tension member. Tension member and gusset plate are both 1/8 in. thick. Determine (a) the total length of the weld required and (b) the length of the weld on each side of the tension member.

6. Solve Sample Problem 11.1 if the following changes are made: The angle member is 5 × 5 × 3/8, and the centroidal axis is 1.39 in. from the edge. The fillet weld has a 1/4 in. leg.

7. (Extra Credit) Solve Problem 6, but also include a 2 in. long end weld placed so that 1 in. lies on either side of the centroidal axis.

8. (SI Units) Two 6061–T6 aluminum tension bars are butt-welded together. They are 75 mm wide and 4 mm thick. What is the allowable load in newtons?
9. (SI Units) A steel bar is lap-welded to a plate, as in Figure 11.1(b). There are two side welds and one end weld. The bar is 40 mm wide by 3 mm thick. The overlap is 100 mm.
   (a) Is the fillet weld stronger than the steel bar in tension?
   (b) What is the allowable tensile load on the joint?

Bolted or Riveted Joints

Unless otherwise specified, use A36 steel and the allowable stress information in the Appendix. Consider hole sizes to be effective hole diameters, and also consider bolt and rivet sizes to be the same as effective hole sizes.

10. A flat steel bar, 3 in. wide by 1 in. thick, is attached by pin to a wall. The hole size for the pin is 1 in. in diameter.
    (a) What allowable tensile load can be applied to the steel bar?
    (b) Use two 1/2 in. diameter bolts side by side, and determine the allowable tensile load. The shank diameters of the bolts take the shear load.
    (c) Use two 1/2 in. rivets, one behind the other (in tandem) along the centroidal axis of the steel bar.
11. Refer to Figure 11.3(a). The plates are 8 in. wide by 1/4 in. thick. Rivet holes are 1 in. in diameter and are spaced equally.
    (a) Determine tensile, bearing, and shear stresses if a tensile load of 25 000 lb is applied.
    (b) Using the allowable stresses, determine the load that can be applied to the joint.
12. A tension member of a bridge is made with two C 5 × 9 channels placed back to back and attached to a gusset plate by four 3/4 in. rivets placed in line along the longitudinal axes of the channels. The gusset plate is 1/2 in. thick and is slipped between the channels.
    (a) What allowable tensile load can the joint carry?
    (b) What is the joint efficiency?
13. Refer to Figure 11.3(b). The plates are 4 in. by 3/8 in., and the rivets are 5/8 in. in diameter.
    (a) What tensile load can the joint carry?
    (b) What is the joint efficiency?

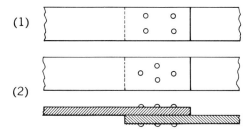

FIGURE 11.5   Problem 14

14. Figure 11.5 shows two different rivet patterns. The plates in each case are 5 in. by 1/4 in., and the rivet holes are 7/8 in. in diameter.
    (a) Determine the load each joint can take.
    (b) Which joint can probably take the largest load if the rivet size is all that can be varied?
15. Refer to Figure 11.3(c). All plates are 10 in. wide by 1/2 in. thick. Rivet holes are 1 in. in diameter.
    (a) What is the allowable tensile load?
    (b) What is the joint efficiency?
16. We are given a double-riveted butt joint with two cover plates. There are four rivets in each main plate. The main plates are 5 in. wide by 7/16 in. thick. Cover plates are 3/8 in. thick. Rivet holes are 1 in. in diameter. Determine the allowable tensile load and the joint efficiency.
17. Refer to Figure 11.3(d). Consider this a random length of a longitudinal seam on a pressure vessel 5 ft in diameter and 25 ft long. Main plates are 1/2 in. thick, cover plates are 7/16 in. thick, and rivet holes are 1 in. in diameter. The short pitch (inner row) is 2½ in.; the long pitch (outer row) is 5 in. *Note:* The length of a repeating section is 5 in.
    (a) What load can one repeating section take?
    (b) What is the joint efficiency?
    (c) What is the allowable pressure in the vessel?
    (d) If the efficiency were 90 percent, what would the pressure be?
18. (Extra Credit)   If the vessel in Problem 17 were 15 ft in diameter, and the internal pressure were 200 psi, determine the stresses in a repeating section for the net area in tension, shear areas, and bearing areas.

216 CHAPTER ELEVEN

19. (Extra Credit) Figure 11.6 shows a W 10 × 100 wide-flange beam connected to a column by means of two L 9 × 4 × 1/2 angles, each 8 in. long. Rivets are 1 in. in diameter. Note carefully the rivet pattern and the shear areas. The beam supports a vertical load, and only a vertical reaction is present. What is the allowable load on the connection?

20. (Extra Credit) Figure 11.7 shows a repeating section of a double-riveted butt joint for a boiler. The top cover plate contains just one row of rivets. Use the dimensions in Problem 17, and determine the tensile load and joint efficiency for one repeating section.

21. (Extra Credit) Determine the probable stresses on a bicycle chain link. You are expected to determine all data needed and take required measurements (or make rational judgments).

22. (SI Units) Solve metric Sample Problem 11.4, but make the rivet holes 25 mm in diameter.

23. (SI Units) A riveted butt joint is composed of two steel main plates. Each is 12 mm thick and 125 mm wide and has one row of three rivets. Each rivet hole diameter is 25 mm. Determine the allowable load the joint can take and its efficiency. Disregard the cover plates.

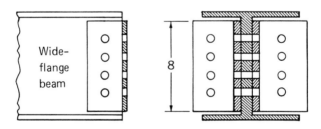

FIGURE 11.6   Problem 19

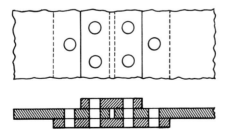

FIGURE 11.7   Problem 20

COMPUTER PROGRAM

The following program will aid in the solution of riveted joint problems. It will solve for the allowable load on a single row. If there is more than one row of rivets in the joint, you will have to account for the other rows to arrive at the correct joint load. Also, you may want to change some of the input data requirements by listing them in the program as constants.

```
10   REM   RIVET
20   PRINT "THIS PROGRAM WILL DETERMINE THE ALLOWABLE LOAD ON A JOINT AT A
       SINGLE ROW."
21   PRINT "IF THERE IS MORE THAN ONE ROW, YOU MUST ADJUST THE ANSWER FOR
       EACH ROW"
22   PRINT "TO ITS CORRECT PERCENT OF THE FULL LOAD.  DATA TO BE IN INCHES
       , POUNDS,AND PSI."
30   PRINT
40   PRINT "TYPE IN ORDER: PLATE WIDTH; PLATE THICKNESS; NUMBER OF RIVET A
       REAS IN SHEAR"
41   PRINT "IN A ROW; DIA. OF RIVET; ALLOWABLE TENSILE, COMPRESSIVE, AND S
       HEAR STRESSES."
50   INPUT W,T,N,D,TS,CS,SS
60   LET PS = SS * 3.1416 * D ^ 2 * N / 4
70   LET PC = CS * N * T * D
80   LET PT = (TS * T * W) - (TS * T * N * D)
90   LET PG = TS * W * T
100  PRINT
110  PRINT "SHEAR LOAD = "; INT (PS * 10 + .51) / 10;" LB."
120  PRINT "BEARING LOAD = "; INT (PC * 10 + .51) / 10;" LB."
130  PRINT "TENSILE LOAD = "; INT (PT * 10 + .51) / 10;" LB."
140  PRINT " GROSS PLATE LOAD = "; INT (PG * 10 + .51) / 10;" LB."
150  PRINT
160  REM   THE FOLLOWING IS A SORT SEQUENCE
170  IF PS < PC THEN   GOTO 200
180  IF PC < PT THEN   GOTO 320
190  GOTO 210
200  IF PS < PT THEN   GOTO 300
210  LET X = PT
220  GOTO 400
300  LET X = PS
310  GOTO 400
320  LET X = PC
330  GOTO 400
400  PRINT "THE ALLOWABLE JOINT LOAD = "; INT (X * 10 + .51) / 10;" LB."
410  PRINT
420  PRINT "THE EFFICIENCY ="; INT ((X / PG) * 1000 + .51) / 10;" %."
430  END
```

12

Torsion on Circular Shafts and Couplings

OBJECTIVES

This chapter develops fundamental torsion formulas so that you will have a deeper and clearer understanding of their application. You will be able to solve for the maximum shear stress in a circular shaft caused by torque and design a shaft to withstand a given torque. You will also be able to determine the angle of twist of a shaft with an applied torque and design a satisfactory bolted flange coupling to transmit a given torque.

INTRODUCTION

Torsion, the act of twisting or turning, is a factor that must be considered in almost all machines and structures. Most machines are driven by an electric motor that supplies a torque; bolts and screws frequently must be tightened to some torque specification; and columns and beams in a structure may have a torque applied about their longitudinal axis because of the type of loading.

NEW SYMBOLS

The following new symbols are used in this chapter and presented here for easy reference:

T = torque in pound·inches

G = modulus of rigidity (modulus of elasticity in shear) in pounds per square inch

J = polar moment of inertia in inches^4 (inches to the fourth power)

ρ (Greek letter *rho*) = radius, in inches, to a given point on cross section of shaft

c = radius in inches to outer fibre of shaft

N = revolutions per minute (rpm) in the horsepower formula

Note that G is listed for various materials in the Appendix. J is also listed in the Appendix for various standard areas.

SHEAR STRESS ON A SHAFT CROSS SECTION

You can easily demonstrate that shear forces develop on cross sections when a torque is applied about the longitudinal axis of a shaft by pressing down and twisting a stack of papers with your hand. One paper will slide past another. Since some force parallel to each page causes the sliding, the force is, by definition, a shear force.

The stresses on small bits of cross section of a shaft make up the total shear force on the cross section but are not uniform. We cannot use the formula $\tau = F/A$ for the solution of torsion problems. To understand why, refer to Figure 12.1, which shows a shaft being subjected

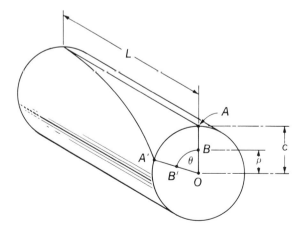

FIGURE 12.1 Strain Relationships on a Shaft Cross Section

TABLE 12.1 Shear Stress in a Circular Shaft

Step	Position at B		Outer Fibre	Comments
1	$\dfrac{B\text{-}B'}{\rho}$	=	$\dfrac{A\text{-}A'}{c}$	
2	$\dfrac{\delta_B}{\rho}$	=	$\dfrac{\delta_A}{c}$	Substitute δ for deformation.
3	$\dfrac{\epsilon_B \cdot L}{\rho}$	=	$\dfrac{\epsilon_A \cdot L}{c}$	Substitute $\epsilon \cdot L$ for δ.
4	$\dfrac{\tau_B \cdot L}{G \cdot \rho}$	=	$\dfrac{\tau_A \cdot L}{G \cdot c}$	From $G = \tau/\epsilon$, substitute τ/G for ϵ.
5	τ_B	=	$\dfrac{\tau_A \cdot G \cdot \rho \cdot L}{G \cdot c \cdot L}$	Rearrange.
6	τ_B	=	$\dfrac{\tau_{max} \cdot \rho}{c}$	Substitute τ_{max} for τ_A. Units: τ in psi; ρ and c in in.

Note: Letter symbols refer to those in Figure 12.1.

to a torque. Point A is twisted to A'. The arc length $A\text{-}A'$ represents the deformation of the outer fibre of a shaft of length L.

Now if we look at a point B somewhere on the radius of the shaft, we note that its deformation is $B\text{-}B'$. It is obvious that the deformation at the outer fibre is greater than at some position closer to the center. (Bear in mind, this assumes that no part of the material exceeds its elastic limit.) Note that the deformations are arcs of circles having the same central angle theta, θ. Therefore, the amount of deformation is directly proportional to the radius.

Shear strains are determined by dividing the deformation by the length of the shaft. The direction of strain is at right angles to the length. This is different from a tensile or compressive strain that takes place in line with the length of the machine part. The relation of shear stress to strain (τ/ϵ) does not give the same modulus of elasticity value obtained with tensile or compressive strains ($E = \sigma/\epsilon$). Instead, another constant is obtained and is called the *shearing modulus* or *modulus of rigidity*. The symbol is G, and the stress–strain relationship is $G = \sigma/\epsilon$. Table 12.1 develops shear stress from strain, and the final step relates shear stress at any point in the shaft cross section to the shear stress at the outer fibre. When a circular shaft has only a torque applied, the maximum shear strain (and, therefore, the maximum shear stress) is at the outer fibre.

TORSION FORMULA 1, RELATING SHEAR STRESS TO TORQUE

The problem now is to relate the shear stress at the outer fibre to the torque applied to the shaft. This relationship is described in Table 12.2 and in Figure 12.2. A twisting action, or moment, applied about the longitudinal axis of a shaft is commonly called *torque*. The shaft may be rotating, as in an electric motor, or it may be stationary, as a bolt. The moment equation from Chapter 3 is repeated here: $M = F \times D$, which is read: *The moment is equal to the force times the perpendicular distance to the moment center.* In this chapter, the symbol M is replaced by the symbol T, for torque.

Refer to Figure 12.2 and Table 12.2 as the torsion formula is developed in the following discussion. Keep in mind that the shear stress varies from zero at the center of the shaft to a maximum at the outer fibre.

Let us consider an extremely small annular area on the cross section of the shaft in Figure 12.2. Its average distance from the moment center, which is the center of the shaft, is indicated by the Greek letter *rho*, ρ. The arrows represent portions of the total force applied to the annular area; they are perpendicular to the radius at their points of application. Let the torque transmitted by this annular area be written as in Step 1 of Table 12.2:

$$T' = f \cdot \rho$$

where

T' (tee prime) = torque on annular area
f = force acting on annular area

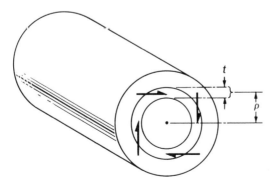

FIGURE 12.2 Diagram of Forces Acting on an Annular Area

TABLE 12.2 Formula 1, Relating Maximum Torsional Shear Stress to Torque

Step	Torque	=	Force × Distance	Comments
1	T'	=	$f \cdot \rho$	T' = torque on an annular area; f = force on annular area.
2	T'	=	$(\tau' \cdot a)\rho$	$\tau' = f/a$; substitute $\tau \cdot a$ for f. τ' = stress on annular area.
3	$T = \Sigma T'$	=	$\Sigma(\tau' \cdot a)\rho$	T = total torque on shaft cross section
4	T	=	$\Sigma \dfrac{\tau_{max} \cdot \rho}{c} \cdot a \cdot \rho$	Substitute Step 6 in Table 12.1 for τ'.
5	T	=	$\dfrac{\tau_{max}}{c} \Sigma \rho^2 \cdot a$	Rearrange terms and remove constants from summation.
6	T	=	$\dfrac{\tau_{max}}{c} \cdot J$	Substitute J for $\Sigma \rho^2 \cdot a$.
	Formula 1: $\tau =$		$\dfrac{T \cdot c}{J}$	Units are: $\dfrac{lb}{in.^2} = \dfrac{(lb \cdot in.) \cdot in.}{in.^4}$

The thickness t of this annular area must be small enough so that any difference in the forces between the inner edge and the outer edge can be neglected. This being the case, the formula $\tau' = f/a$ may be used to relate the total force on the annular area to the stress. Substituting $\tau' \cdot a$ for f in the torque formula, as in Step 2 of Table 12.2:

$$T' = (\tau' \cdot a)\rho$$

where

τ' (tau prime) = shear stress on annular area
a = area of the very small annular ring

What is needed is the sum of all the torques on all of the annular areas in the cross section. Therefore, as in Step 3 of Table 12.2:

$$T = \Sigma T' = \Sigma(\tau' \cdot a)\rho$$

where

T = total torque on cross-sectional area of shaft

Before proceeding with the summation, one other algebraic step must be performed. It is desirable to have an expression for torque using only the maximum stress appearing on the outer fibre of the shaft. By substituting values from Step 6 in Table 12.1, the torsion formula is as shown in Step 4 of Table 12.2:

$$T = \sum \frac{\tau_{max} \cdot \rho}{c} \cdot a \cdot \rho$$

In this instance, τ' is considered to be the same as τ_B.

Constants are factored out of the summation, and the formula is rearranged as shown in Step 5 of Table 12.2:

$$T = \frac{\tau_{max}}{c} \Sigma \rho^2 \cdot a$$

The summation of $\rho^2 \cdot a$ is called the *polar moment of inertia* and is designated J for convenience. Substitution of J gives the formula in Step 6 of Table 12.2:

$$T = \frac{\tau_{max}}{c} \cdot J$$

The formula is rearranged and the subscript is dropped (because τ is understood to be the maximum torsional shear stress) to obtain the most commonly written form:

Formula 1: $\quad \tau = \dfrac{T \cdot c}{J}$

The formula reads:

The stress on the outer fibre of a shaft is equal to the torque multiplied by the shaft radius and divided by the polar moment of inertia.

For a solid circular shaft,

$$J = \frac{\pi D^4}{32}$$

For a hollow circular shaft,

$$J = \frac{\pi(D^4 - d^4)}{32}$$

where D is the outside diameter and d is the inside diameter.

Sample Problem 12.1 and metric Sample Problem 12.2 illustrate the application of Torsion Formula 1.

SAMPLE PROBLEM 12.1
Torsional Shear Stresses Developed in a Shaft

Problem

Given the two steel shafts shown, each having a torque of 1000 lb·in. applied to it, and with cross-sectional areas equal, determine (a) the stress on the outer fibre of shaft A, (b) the stress on the outer fibre of shaft B, (c) the stress 1/8 in. below the surface of shaft A, and (d) the stress on the inner surface of shaft B.

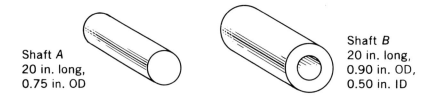

Shaft A
20 in. long,
0.75 in. OD

Shaft B
20 in. long,
0.90 in. OD,
0.50 in. ID

Solution

$$J_A = \frac{\pi D^4}{32} = \frac{\pi (0.75)^4}{32} = \frac{\pi}{32}(0.316) = 0.031 \text{ in.}^4$$

$$J_B = \frac{\pi (D^4 - d^4)}{32} = \frac{\pi}{32}[(0.9)^4 - (0.5)^4]$$

$$= \frac{\pi}{32}(0.656 - 0.063) = 0.058 \text{ in.}^4$$

Answer (a)

$$\tau = \frac{T \cdot c}{J} = \frac{1000 \cdot (0.75/2)}{0.031} = 12\,100 \text{ psi}$$

Answer (b)

$$\tau = \frac{T \cdot c}{J} = \frac{1000 \cdot (0.90/2)}{0.058} = 7760 \text{ psi}$$

Answer (c)

Stresses are proportional to their distances from the shaft axis:

$$\frac{12\,100}{0.75/2} = \frac{X\text{ psi}}{0.25}$$

$$X = 8070 \text{ psi}$$

Answer (d)

$$\frac{7760}{0.45} = \frac{X\text{ psi}}{0.25}$$

$$X = 4310 \text{ psi}$$

METRIC SAMPLE PROBLEM 12.2
Torsional Shear Stress

Problem

Find the torsional shear stress in pascals on the outer fibre of a steel shaft 40 mm in diameter, with an applied torque of 75 N·m. Note that since stress is in pascals (newtons/metre^2), dimensions in the formula must be in metres.

Solution

Since $\tau = (T \cdot c)/J$ and $J = \pi D^4/32$, we have

$$\tau = \frac{75\,(0.04/2)}{[\pi(0.04)^4]/32} = 5.95(10)^6 \text{ Pa} = 5.95 \text{ MPa}$$

TORSION FORMULA 2, RELATING ANGLE OF TWIST TO STRESS

The shaft of a generator must have a twist limitation, or opposite ends of the armature will pass their pole positions slightly out of phase. Shafts operating through timing gears in an engine, torsion bar suspension, and shafts geared to each other in precision equipment must have a designated amount of twist. This limitation is the allowable angle of twist theta, θ, shown in Figure 12.1. You must be prepared for requirements such as "the shaft must not have an angle of twist greater than 2° in ten diameters of length."

TABLE 12.3 Formula 2, Relating Angle of Twist to Stress

Step	Angle θ = $\dfrac{\text{Arc Length}}{\text{Radius}}$		Comments
1	θ =	$\dfrac{\delta}{c}$	θ is in radians. $\delta = \epsilon \cdot L$; make substitution in Step 2.
2	θ =	$\dfrac{\epsilon \cdot L}{c}$	$G = \tau/\epsilon$; rearrange and substitute τ/G for ϵ.
Formula 2:	θ =	$\dfrac{\tau \cdot L}{G \cdot c}$	Units are: radians = $\dfrac{(\text{lb/in.}^2)(\text{in.})}{(\text{lb/in.}^2)(\text{in.})}$

The derivation of the formula relating the angle of twist to the stress in the outer fibre is given in Table 12.3. This derivation is based on three formulas with which you are already familiar:

1. The formula θ = arc length/radius comes from trigonometry.
2. The formula for strain is $\epsilon = \delta/L$. δ is the deformation of the outer fibre represented by A-A' in Figure 12.1.
3. The formula $G = \tau/\epsilon$ relates the modulus of rigidity to stress and strain.

These formulas are combined in Table 12.3 to give us Torsion Formula 2:

$$\text{Formula 2:} \quad \theta = \frac{\tau \cdot L}{G \cdot c}$$

This formula reads:

The angle of twist (radians) is equal to the stress in the outer fibre, times the length (in.) divided by the modulus of rigidity (psi), times the radius of the outer fibre (in.).

Another useful formula is

$$\text{Formula 3:} \quad \theta = \frac{T \cdot L}{J \cdot G}$$

This formula reads:

The angle of twist is equal to the torque (lb·in.), times the length of the shaft (in.), divided by the polar moment of inertia (in.^{4}), times the modulus of rigidity.

It relates angle of twist to torque and is derived from Torsion Formulas 1 and 2. The derivation is left as an exercise in Problem 15 at the end of this chapter.

Formulas 2 and 3 are based on two assumptions: (1) that the shaft is in equilibrium, which means the shaft is not accelerating but is either stationary or is rotating at a constant speed, and (2) that the elastic limit for the material is not exceeded.

When considering rotating shaft problems, the formula relating horsepower to torque is

Formula 4: $\quad \text{hp} = \dfrac{T \cdot N}{63\,000}$

where
T = torque in lb·in.
N = rpm

Sample Problem 12.3 illustrates an angle of twist problem and its solution.

SAMPLE PROBLEM 12.3
Angle of Twist

Problem

A 1/8 in. diameter shaft has an allowable shear stress of 13 500 psi. The shaft is brass. (a) How long must the shaft be to develop a full 360° twist? (b) If the maximum allowable twist is 1° in 30 diameters, what shear stress is developed on the outer fibre? (c) If condition (b) applies, and the shaft rotates at 8000 rpm, what hp is transmitted?

Solution

$G = 6(10)^6$ psi (from Appendix)

$J = \dfrac{\pi D^4}{32} = \dfrac{\pi (1/8)^4}{32} = 23.95(10)^{-6}$ in.^{4}

Answer (a)

$\theta = 360° = 2\pi \text{ radians} = \dfrac{T \cdot L}{G \cdot c}$

$L = \dfrac{\theta \cdot G \cdot c}{T} = \dfrac{2\pi \cdot 6(10)^6 \cdot 1/16}{13\,500} = 174$ in. $= 14.5$ ft

Answer (b)

$$\frac{X \text{ radians}}{1°} = \frac{2\pi \text{ radians}}{360°}$$

$1° = 0.0174$ radians

$$\tau = \frac{\theta \cdot G \cdot c}{L} = \frac{0.0174 \cdot 6(10)^6 \cdot 1/16}{30(1/8)}$$

$$= \frac{174(10)^{-4} \cdot 6(10)^6}{(30/8) \cdot 16} = 1740 \text{ psi}$$

number of diameters

Answer (c)

$$hp = \frac{T \cdot N}{63\,000}$$

$$T = \frac{\theta \cdot J \cdot G}{L}$$

$$hp = \frac{\theta \cdot J \cdot G \cdot N}{63\,000 \cdot L} = \frac{174(10)^{-4} \cdot 23.95(10)^{-6} \cdot 6(10)^6 \cdot 8000}{63\,000(30/8)} = 0.084$$

COUPLINGS

Coupling problems require that the student keep in mind that a shaft and/or coupling arrangement transmits torque and *not* a force. For instance, a force applied at the outer fibre of a 3 in. shaft is not the same force appearing on the bolts of a coupling with a bolt circle diameter of 10 in. However, the torque transmitted by the shaft is the same as the torque transmitted by the bolts in the coupling. Generally speaking, the solution of a coupling problem requires the determination of the torque involved. The next step is to visualize a single bolt used to transmit torque through the coupling. This bolt would have a force applied to it sufficient to transmit the torque. This force is then used to help determine the number of bolts required in the coupling, as illustrated in Sample Problem 12.4.

One must be careful to apply Torsion Formula 1 to the shaft but not to the bolts. The formula $\tau = F/A$ can be used for the shear stress in the bolts because the bolts are not twisted about their own axes but have a direct force applied.

SAMPLE PROBLEM 12.4
Coupling

Problem

The coupling transmits 70 hp at 200 rpm. All materials are steel with an allowable shear stress of 15 000 psi. Note that the shear area on each bolt is at the shank diameter of the bolt and not at the minor thread diameter. (a) If the shear stress on the shaft is above allowable, specify shaft size. (b) Four bolts are shown. If this is not sufficient, specify number. (c) If the shafts rotate at 400 rpm, will more or fewer bolts be required? Horsepower is to remain constant.

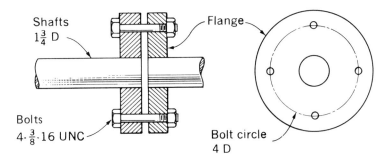

Solution (a)

Horsepower must be converted to torque so that Torsion Formula 1 can be used.

$$T = \frac{hp \cdot 63\,000}{N} = \frac{70 \cdot 63(10)^3}{2(10)^2} = 22\,100 \text{ lb·in.}$$

Using allowable stress, the required shaft diameter is calculated:

$$\tau = \frac{T \cdot c}{J}$$

$$\frac{J}{c} = \frac{T}{\tau}$$

$$\frac{\pi D^4}{32(D/2)} = \frac{22\,100}{15\,000}$$

$$D^3 = \frac{16 \cdot 22\,100}{\pi \cdot 15\,000} = 7.50 \text{ in.}^3$$

$$D = \sqrt[3]{7.50} = 1.96 \text{ in. (minimum diameter)}$$

Therefore, shaft size given is too small. Increase diameters to 1.96 in.

Solution (b)

If we imagined all the torque transmitted by one bolt, the FBD of one flange would look like the following:

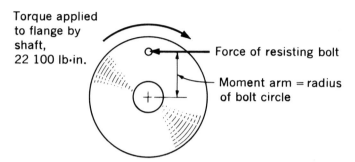

Torque applied to flange by shaft, 22 100 lb·in.

Force of resisting bolt

Moment arm = radius of bolt circle

$T = F \cdot d$

$F = \dfrac{T}{d} = \dfrac{22\,100}{2 \text{ in.}} = 11\,050$ lb applied to bolts

$\tau = \dfrac{F}{A}$

$A = \dfrac{F}{\tau} = \dfrac{11\,050}{15\,000} = 0.736$ in.^{2} (total area required)

Note that 3/8D bolts are specified. Since the shank area of one 3/8 bolt = 0.110 in.^{2},

number of bolts required $= \dfrac{0.736}{0.110} = 6.7 = 7$ bolts (rounded)

Solution (c)

From the equation $T = (\text{hp} \cdot 63\,000)/N$, it can be seen that torque varies inversely with N. Therefore, if hp is constant, torque will be reduced, and the number of bolts can be reduced. On the other hand, if torque is maintained constant, the shafts and 7 bolts can transmit greater hp at the higher speed.

LONGITUDINAL SHEAR FORCE

You should be aware that an applied torque on a shaft develops a longitudinal shear force in the shaft. Let us look at a thin disc cut from

Torsion on Circular Shafts and Couplings   231

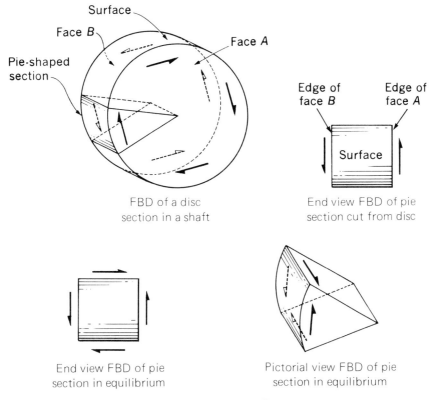

FIGURE 12.3   Longitudinal Shear Forces

a shaft, as shown in Figure 12.3. Arrows indicate the force applied to each cross-sectional surface. If a section is cut from this disc, and a free body diagram drawn, it can be noted that shear forces are developed in opposite directions on the cross-sectional faces A and B, producing rotation of the section. For the section to remain in equilibrium, longitudinal shear forces must also be present for the summation of moments to be zero. You will also recognize that the shear forces on the longitudinal faces of the section must equal the shear forces on the cross-sectional faces. A shaft that is weaker in shear along its axis may fail in a longitudinal direction; wooden shafts formerly used in some chemical processes fall into this category. Also, forged axles in cars and trucks sometimes fail in a longitudinal direction because forging tends to strengthen the axle more in one direction than another.

PHOTO 12.1   Gears for Use in Marine Propulsion Systems at a General Electric Plant

SUMMARY

The fundamental torsion formulas for circular shafts are as follows:

$$\text{Formula 1:} \quad \tau = \frac{T \cdot c}{J}$$

$$\text{Formula 2:} \quad \theta = \frac{\tau \cdot L}{G \cdot c}$$

$$\text{Formula 3:} \quad \theta = \frac{T \cdot L}{J \cdot G}$$

$$\text{Formula 4:} \quad \text{hp} = \frac{T \cdot N}{63\,000}$$

The polar moment of inertia for solid circular shafts is:

$$J = \frac{\pi D^4}{32}$$

The polar moment of inertia for hollow circular shafts is:

$$J = \frac{\pi(D^4 - d^4)}{32}$$

PROBLEMS

Shear Stress

1. A 3 in. OD solid steel shaft has an external shear stress of 9000 psi. What is the shear stress (a) 1 in. from the shaft center, (b) 3/4 in. from the center, and (c) at the center?
2. A hollow steel shaft has a 6 in. OD and a 3 in. ID. The shear stress on the outer fibre of the shaft is 5000 psi. What is the shear stress (a) on the internal surface and (b) 1 in. from the center of the shaft?
3. (SI Units) Refer to Figure 12.1. Let radius OB equal 9.5 mm, radius OA equal 13 mm, and the shear stress at B equal 22 MPa. What is the maximum shear stress on the outer fibre of the shaft?

Torsion Formula 1

4. Repeat Sample Problem 12.1, but double all of the diameters.
5. Figure 12.4 shows a four-way steel wheel wrench. Assume the

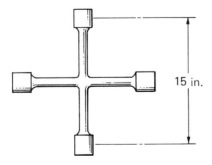

FIGURE 12.4   Problem 5

maximum force a person can apply with each hand when turning the wrench is approximately 100 lb with the hands 15 in. apart. The diameter of each shaft is 7/8 in. Assuming the stress is below the elastic limit for the steel, what torsional shear stress occurs on the shaft being twisted?

6. Figure 12.5 shows a standard steel torque wrench, with a capacity of 100 lb·ft. The connector between the end of the wrench and the socket is 1/2 in. in diameter. What torsional shear stress develops in this connector?
7. Specify the diameter of a solid steel shaft (allowable shear stress equals 15 000 psi) that must take a 10 000 lb·in. torque.
8. A hollow steel shaft has a 3 in. OD and a 1.75 in. ID. A 12 000 psi torsional shear stress is developed on the outer fibre. Determine (a) the shear stress developed on the inner surface and (b) what torque is applied.
9. Repeat Problem 7 for a hollow shaft with an outside diameter of 3 in. Specify the inside diameter.
10. A 5/8 in. diameter solid steel shaft 15 in. long has a drive pulley at the left end and two take-off pulleys placed at the middle and at the right end of the shaft. The drive pulley applies a 300 lb·in. torque, the middle pulley takes off 150 lb·in., and the end pulley takes off the remaining torque. What is the maximum shear stress on each half of the shaft?
11. (SI Units) A 90 N·m torque is applied to an aluminum shaft 12 mm in diameter. Calculate the maximum shear stress.

Angle of Twist

12. Solve Sample Problem 12.3, but consider the shaft material to be steel.
13. What is the angle of twist on the shaft of the four-way wrench in Problem 5?
14. A hollow steel oil well drilling rod has a 6 in. OD and a 4 in. ID. In drilling, it twists through 720° in 2000 ft. Determine (a) what torque is applied and (b) what torsional shear stress develops on the outer fibre.
15. (Extra Credit) Derive the formula $\theta = (T \cdot L)/(J \cdot G)$ from $\tau = (T \cdot c)/J$ and $\theta = (\tau \cdot L)/(G \cdot c)$.
16. (Extra Credit) Calculate the total angle of twist in Problem 10.
17. (Extra Credit) An aluminum shaft with two diameters is shown in Figure 12.6. The shaft has a 2000 lb·in. torque applied at the right end; the wall supplies the resisting torque. Determine

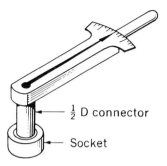

FIGURE 12.5   Problem 6

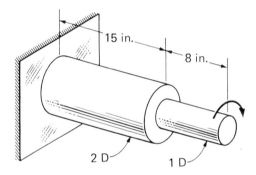

FIGURE 12.6   Problem 17

(a) the total angle of twist for the 23 in. length, and (b) the torsional shear stress on the outer fibre of each section of the shaft.

18. (SI Units)   In Sample Problem 12.2, what is the angle of twist if the shaft is 1.5 m long?

Couplings

19. A diesel electric locomotive has an 1800 hp engine connected to a generator.
    (a) If the drive shaft of the engine is steel, 3 in. OD, and rotates at 1200 rpm, what torque is developed?
    (b) If the shaft is coupled to the generator with eight 3/4 in. bolts in a coupling having a 5 in. diameter bolt circle, what shear stress is developed in each bolt?
    (c) Solve (a) and (b) if the shaft rotates at 2400 rpm. The horsepower remains the same.

20. The American Bureau of Shipping gives the following formula for determining the required diameter of steel line shafting (connecting propeller shaft to the turbine shaft):

$$D = \sqrt[3]{\frac{K \cdot hp}{N}}$$

where
D = diameter in inches
K = 64 for ocean service
N = rpm

A cargo ship line shaft operates at 105 rpm and 4500 hp.
(a) Determine the shaft size with the formula above.
(b) What torsional shear stress is developed in the outer fibre of the solid shaft?
(c) Assume the coupling has an 18 in. diameter bolt circle, 2 in. diameter bolts are used, and allowable shear stress for the bolts is 13 000 psi. Determine the number of bolts required.

21. (SI Units) What torque can the following coupling transmit at 300 rpm? The coupling consists of a 150 mm bolt circle diameter and four bolts with diameters of 15 mm; the bolt material has an allowable stress of 70 MPa.

COMPUTER PROGRAM

The following program will help you solve for shear stress, torque, angle of twist, and horsepower. If there are some answers you do not need, delete the appropriate part of the program.

```
10  REM  TORSION
20  PRINT "THIS PROGRAM WILL SOLVE FOR EITHER THE SHEAR STRESS ON THE OUTER "
21  PRINT "CIRCUMFERENCE OR THE TORQUE ON A SHAFT AND THEN THE ANGLE OF TWIST AND HORSEPOWER."
30  PRINT "IF YOU WANT TO SOLVE FOR SHEAR STRESS - TYPE 1, FOR TORQUE - 2 ."
40  INPUT A
50  IF A = 1 THEN  GOTO 70
60  IF A = 2 THEN  GOTO 120
70  PRINT "TYPE IN ORDER: TORQUE (LB.IN.), OUTER DIA. (IN.), INNER DIAMETER "
71  PRINT "(TYPE - 0 - IF SHAFT IS SOLID), LENGTH OF SHAFT (IN.), RPM, AND 'G' (PSI."
```

```
80   INPUT T,DO,DI,L,N,G
90   LET S = (16 * T * DO) / (3.1416 * ((DO ^ 4) - (DI ^ 4)))
95   PRINT
100  PRINT "SHEAR STRESS = "; INT (S * 10 + .51) / 10;" PSI."
110  GOTO 160
120  PRINT "TYPE IN ORDER: SHEAR STRESS (PSI, OUTER DIA. (IN.), INNER DIA
     METER "
121  PRINT "(TYPE - 0 - IF SHAFT IS SOLID), LENGTH OF SHAFT (IN.), RPM, A
     ND 'G' (PSI."
130  INPUT S,DO,DI,L,N,G
140  LET T = S * 3.1416 * ((DO ^ 4) - (DI ^ 4)) / (16 * DO)
145  PRINT
150  PRINT "TORQUE = "; INT (T * 10 + .51) / 10;" LB.IN."
160  PRINT
170  LET ANG = S * (L / G) * (2 / DO) * (57.296)
180  PRINT "ANGLE OF TWIST = "; INT (ANG * 100 + .51) / 100;" DEG."
190  PRINT
200  LET HP = T * N / 63000
210  PRINT "POWER = "; INT (HP * 100 + .51) / 100;" HP."
220  END
```

Shear and Moment Diagrams

OBJECTIVES

This chapter explains shear and moment diagrams so that you can construct your own diagrams from information about beam loads. It also provides the initial steps for the solution of beam problems.

INTRODUCTION

A *beam* is defined as a member of a structure or machine that carries a load applied at an angle to the beam's main longitudinal axis. There are two general classes of beams: *statically determinate* and *statically indeterminate*. Figure 13.1 depicts several types of loading and support in both classes.

The reactions of statically determinate beams can be solved using only the statics equilibrium equations: $\Sigma M = 0$, $\Sigma F_x = 0$, and $\Sigma F_y = 0$. Statically indeterminate beams require additional formulas for solution; some are considered in Chapter 18.

To solve beam problems, it is necessary to understand the complete force systems applied to a beam and the manner in which these forces are resisted internally in the beam. Figure 13.2(a) illustrates a cantilever beam supporting a hanging load. If a hinge should be put in the middle of the beam, it is obvious that the end of the beam will swing down to a new equilibrium position, as shown in Figure 13.2(b). In analyzing the loading, we note that the hinge can resist the direct load W but cannot resist an applied moment. The applied moment about the hinge is equal to $W \cdot L_1$. Since the original beam does maintain its

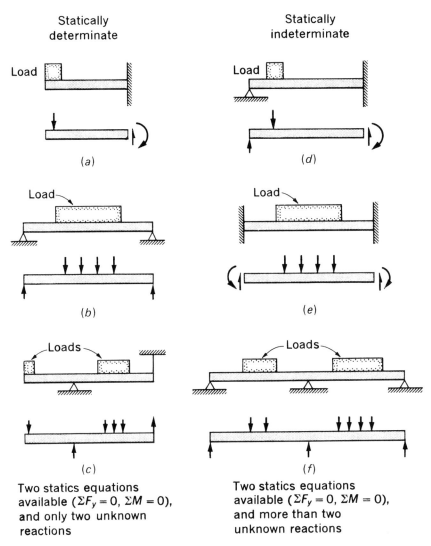

FIGURE 13.1 Classes of Beams

Under (c): Two statics equations available ($\Sigma F_y = 0$, $\Sigma M = 0$), and only two unknown reactions

Under (f): Two statics equations available ($\Sigma F_y = 0$, $\Sigma M = 0$), and more than two unknown reactions

position, it is resisting both the direct load and the moment being produced by the load.

When a beam maintains its equilibrium, every particle in the beam and any given portion of the beam must be in equilibrium. A free body diagram of the left-hand portion of the beam is shown in Figure 13.2(c). For static equilibrium of the left portion of the beam, the applied force W must be resisted by an equal and opposite force and a resisting

240 CHAPTER THIRTEEN

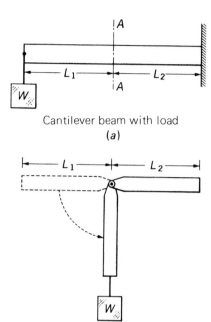

Cantilever beam with load
(a)

Loaded cantilever beam with hinge
(b)

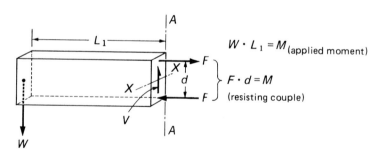

FBD of left portion of beam shown in (a)
(c)

FIGURE 13.2 Applied Forces and Moments on a Beam Cross-Section

couple acting on cross section A-A. The resisting shear force is labeled V in Figure 13.2(c).

The statics equations above can best be applied to beam problems by keeping in mind the following relationships:

$$\Sigma F = 0 = +\Sigma_{\text{applied forces}} - \text{internal resisting shear force}$$
$$\Sigma M = 0 = +\Sigma_{\text{applied moments}} - \text{internal resisting couple}$$

The plus and minus signs indicate only that the terms must be of opposite sign. The moment equation above refers to moments taken about a moment center on the X-X axis of cross section A-A in Figure 13.2(c).

The technician must know what the applied force and applied moment is at all cross sections to design the beam properly. For long structural beams, these calculations may be made at every foot of length. You should remember that, except for finding reactions at the beam supports, the complete statics equilibrium equation is not solved. Only the applied forces and applied moments are calculated; this information is plotted on graphs called *shear diagrams* and *moment diagrams*.

The shear diagram is simply a plot of the applied forces at every point along the length of the beam. These applied forces are given the symbol V because they must be opposed by the internal shear forces. Often, the shear diagram is simply identified by the letter V.

The applied moment at any given location is represented by the letter M. A moment diagram is a plot of the applied moments at every point along the beam. The moment diagram may be identified by the letter M.

SHEAR DIAGRAMS

Sample Problems 13.1 and 13.2 show that before constructing the shear and moment diagrams, you should draw a sketch of the setup to some convenient scale. For instance, for a 25 ft beam, a scale of 1 in. = 10 ft would yield a drawing 2.5 in. long. Place the free body diagram directly under the sketch, and construct the shear and moment diagrams directly under the free body diagram. Plot the applied shear force graphically at each selected cross section of the beam, with the shear force shown on the ordinate (Y-axis) and the distance of the cross section from the left end of the beam plotted on the abscissa (X-axis). The procedure for determining the applied shear force and constructing the shear diagram is given next.

Procedure for Constructing Shear Diagrams

Step 1 Solve for the beam reactions.

Step 2 It is generally accepted practice to determine the shear force values starting at the left of the beam and moving to the right.

Step 3 At each position, make an imaginary cut in the beam and draw

242   CHAPTER THIRTEEN

a free body diagram (if required for clear understanding) of the left portion of the beam.

Step 4   Sum all the applied forces to the left of the cut (cross section). Reactions are considered to be applied forces.

Step 5   At a concentrated load, the practice is to make a cut just to the left and just to the right of the load.

Step 6   In order to provide consistent diagrams for cantilever beams, the beams should be viewed so that the wall is on the right side of the beam. Unless otherwise specified, the weight of the beam is considered negligible and therefore disregarded.

In Sample Problems 13.1 and 13.2, shear values are calculated for every foot of the beam. The subscripts to the shear symbol indicate the position on the beam in feet. For instance, V_3 means that an imaginary cut was made 3 ft from the left end of the beam. The externally applied forces to this 3 ft section of beam are listed and also plotted on the shear diagram.

SAMPLE PROBLEM 13.1
Cantilever Beam with Concentrated Loads

Problem

Given the loaded cantilever beam shown in the top sketch, (a) draw the FBD and determine the reactions, (b) calculate the applied shearing force at every foot of length and plot the shear diagram, and (c) calculate the applied moment at every foot of length and plot the moment diagram.

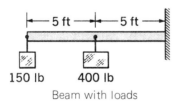
Beam with loads

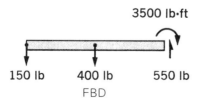
FBD

Shear and Moment Diagrams

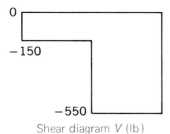
Shear diagram V (lb)

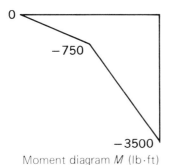
Moment diagram M (lb·ft)

Solution

Solve for the beam reactions:

V (at wall) $= 550$ lb ↑

M (at wall) $= +3500$ lb·ft

Determine the shear force values. *Note*: Read V_0^- as the applied shear force just to the left of the zero foot mark.

$V_0^- = 0$ lb
$V_0^+ = -150$ lb
$V_1 = -150$ lb
$V_{2,3,4} = -150$ lb
$V_5^- = -150$ lb
$V_5^+ = -550$ lb
$V_{6,7,8,9} = -550$ lb
$V_{10}^- = -550$ lb
$V_{10}^+ = 0$ lb

Calculate the applied moments. *Note:* Read M_0 as the moment applied about a moment center at the zero foot mark.

244 CHAPTER THIRTEEN

$M_0 = 0$ lb·ft
$M_1 = -150(1) = -150$ lb·ft
$M_2 = -150(2) = -300$ lb·ft
$M_3 = -150(3) = -450$ lb·ft
$M_4 = -150(4) = -600$ lb·ft
$M_5 = -150(5) = -750$ lb·ft
$M_6 = -150(6) - 400(1) = -1300$ lb·ft
$M_7 = -150(7) - 400(2) = -1850$ lb·ft
$M_8 = -150(8) - 400(3) = -2400$ lb·ft
$M_9 = -150(9) - 400(4) = -2950$ lb·ft
$M_{10} = -150(10) - 400(5) = -3500$ lb·ft
$M_{10}^+ = -3500 + 3500 = 0$

Construct the shear and moment diagrams from the calculated data.

SAMPLE PROBLEM 13.2
Overhanging Beam with Distributed Load and Concentrated Load

Problem

Given the loaded overhanging beam shown in the top sketch, (a) draw the FBD and determine the reactions, (b) calculate the applied shearing force at every foot of length and plot the shear diagram, and (c) calculate the applied moment at every foot of length and plot the moment diagram.

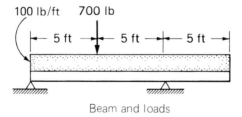
Beam and loads

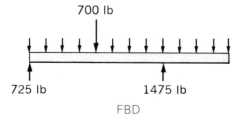
FBD

Shear and Moment Diagrams   245

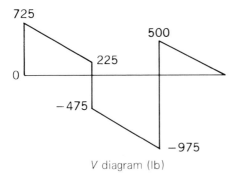

V diagram (lb)

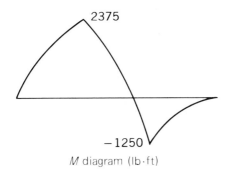

M diagram (lb·ft)

Solution

Determine the reactions. Then construct the graphs from the calculated data.

$V_0 = 0$ lb
$V_0^+ = +725$ lb
$V_1 = +725 - 100 = +625$ lb
$V_2 = +725 - 200 = +525$ lb
$V_3 = +725 - 300 = +425$ lb
$V_4 = +725 - 400 = +325$ lb
$V_5 = +725 - 500 = +225$ lb
$V_5^+ = +725 - 500 - 700 = -475$ lb
$V_6 = +725 - 700 - 600 = -575$ lb
$V_7 = +725 - 700 - 700 = -675$ lb
$V_8 = +725 - 700 - 800 = -775$ lb
$V_9 = +725 - 700 - 900 = -875$ lb
$V_{10} = +725 - 700 - 1000 = -975$ lb
$V_{10}^+ = +725 - 700 - 1000 + 1475 = +500$ lb

$V_{11} = +725 - 700 + 1475 - 1100 = +400$ lb
$V_{12} = +725 - 700 + 1475 - 1200 = +300$ lb
$V_{13} = +725 - 700 + 1475 - 1300 = +200$ lb
$V_{14} = +725 - 700 + 1475 - 1400 = +100$ lb
$V_{15} = +725 - 700 + 1475 - 1500 = 0$ lb

$M_0 = 0$ lb·ft
$M_1 = +725(1) - 100(0.5) = +625$ lb·ft
$M_2 = +725(2) - 200(1) = +1250$ lb·ft
$M_3 = +725(3) - 300(1.5) = +1725$ lb·ft
$M_4 = +725(4) - 400(2.0) = +2100$ lb·ft
$M_5 = +725(5) - 500(2.5) = +2375$ lb·ft
$M_6 = +725(6) - 600(3) - 700(1) = +1850$ lb·ft
$M_7 = +725(7) - 700(3.5) - 700(2) = +1175$ lb·ft
$M_8 = +725(8) - 800(4) - 700(3) = +500$ lb·ft
$M_9 = +725(9) - 900(4.5) - 700(4) = -325$ lb·ft
$M_{10} = +725(10) - 1000(5) - 700(5) = -1250$ lb·ft
$M_{11} = +725(11) - 1100(5.5) - 700(6) + 1475(1) = -800$ lb·ft
$M_{12} = +725(12) - 1200(6) - 700(7) + 1475(2) = -450$ lb·ft
$M_{13} = +725(13) - 1300(6.5) - 700(8) + 1475(3) = -200$ lb·ft
$M_{14} = +725(14) - 1400(7) - 700(9) + 1475(4) = -50$ lb·ft
$M_{15} = +725(15) - 1500(7.5) - 700(10) + 1475(5) = 0$ lb·ft

MOMENT DIAGRAMS

The moment diagram is generally constructed after the shear diagram. The applied moment at each selected cross section is plotted graphically, with the length of the beam plotted along the X-axis and the moment value plotted along the Y-axis. Sample Problems 13.1 and 13.2 illustrate the construction of moment diagrams. The procedure for determining the applied moment at any location on the beam follows.

Procedure for Constructing Moment Diagrams

Step 1 Starting at the left end and moving progressively to the right, make imaginary cuts through the beam, and draw a free body diagram of the portion of the beam to the left of the cut. The cuts are usually made at each 1 ft interval.

Step 2 With the X-X axis of the cross section of the cut as the moment center axis, sum the applied moments to the left of the cut. Plot this on the moment diagram.

Step 3 To provide consistent diagrams for cantilever beams, the beams should be viewed so that the wall is on the right.

RELATIONSHIPS BETWEEN SHEAR AND MOMENT DIAGRAMS

Observe that when the beam has only concentrated loads (Sample Problem 13.1), the shear diagram will consist of a straight horizontal line between loads and straight vertical lines at the concentrated loads. If a distributed load is on the beam (Sample Problem 13.2), the shear diagram has a slanted straight line and the moment diagram has a curved line.

Whenever the shear diagram crosses the zero axis, the moment diagram is a *maximum*. That is, the magnitude (either + or −) is greater than the magnitude just to either side of the maximum position.* Sample Problem 13.3 and metric Sample Problem 13.4 show how to use the shear diagram to determine the position of the maximum moment value.

Another useful relationship is that the area of the shear diagram, up to a given position on the beam, is equal to the moment on the beam at that position. Therefore, once you become well grounded in moment diagrams, you may speed up your work by first drawing the shear diagram and then using it to aid in the drawing of the moment diagram. In such a situation, you calculate only the points of change and the maximum points on the moment diagram and then sketch the diagram to show its shape. For example, in Sample Problem 13.3, there are no applied loads on the left side of the beam, so you plot the shear value at the left end and again at the 10 ft mark. A straight horizontal line connects these two points. A slanted line connects the 10 ft value in the shear diagram with the 20 ft value. To determine the point where the shear diagram crosses the zero axis (the X-axis) and therefore the location of the maximum moment, two methods may be used.

In one method, a simple proportion may be set up for the slanted line of the shear diagram. Because of the distributed load, we know

---

*This statement can be confusing, because in some situations the moment value may actually be zero (as is the case in Figure 13.15).

that the shear value changes at the rate of 100 lb for every foot of length. Also, by observing the shear diagram, we note that it changes from 0 to 250 lb in X ft. The proportion becomes

$$\frac{X \text{ ft}}{250 \text{ lb}} = \frac{1 \text{ ft}}{100 \text{ lb}}$$

Solving the proportion, X is found to be 2.5 ft from the *left end of the distributed load*.

The second method requires writing the equation for the slanted line in the shear diagram. First, we choose a specific distance and write the shear equation for that distance. For example, we will choose a distance of 15 ft from the left end of the beam in Sample Problem 13.3:

$$V_{15} = +250 \text{ lb} - 100 \text{ lb/ft} \times (15 \text{ ft} - 10 \text{ ft})$$
$$= +250 - 100(15 - 10)$$

Next, we rewrite the equation and substitute X for the location's distance (15 ft) from the left end. The equation now becomes

$$V_X = +250 - 100(X - 10)$$

This equation will provide us with the value for any location on the shear diagram from 10 ft to 20 ft. However, our immediate problem is to find the location where $V_X = 0$. Therefore, we set the equation to zero:

$$V_X = 0 = +250 - 100(X - 10)$$

X is found to be 12.5 ft from the *left end of the beam*.

Note: Although each method measures X from a different point on the beam, the methods do agree on the location of the crossover point.

Writing equations for moment diagrams may be useful when distributed loads cover the full length of the beam, as in Sample Problem 13.4. The following suggested procedure is very similar to the one suggested for writing shear diagram equations. The procedure is illustrated using the moment diagram in Sample Problem 13.3.

Procedure for Writing Moment Diagram Equations

Step 1 Choose a specific distance, and write the moment equation for that distance. Example: Refer to Sample Problem 13.3. A 5 ft distance is chosen. The equation is:

$M_5 = 250 \text{ lb} \times 5 \text{ ft} = 250 \cdot 5$

Step 2 Rewrite the equation, and substitute X for the distance used (5 ft):

$M_X = 250 \cdot X$

This general equation for the first moment curve is good for beam distances from 0 ft to 10 ft, measured from the left end of the beam.

The equation for the second curve is a bit more involved. Arbitrarily, we choose to sum the applied moments at the 15 ft section:

$M_{15} = +250 \text{ lb}(15 \text{ ft})$

$\quad\quad\quad - 100 \text{ lb/ft}(15 \text{ ft} - 10 \text{ ft})\left(\dfrac{15 \text{ ft} - 10 \text{ ft}}{2}\right)$

Now X is substituted for 15 ft:

$M_X = +250X - 100(X - 10)\left(\dfrac{X - 10}{2}\right)$

$\quad\quad = +250X - 50X_2 + 1000X - 5000$

$\quad\quad = -50X^2 + 1250X - 5000$

Notice that if the quadratic equation is solved for its roots, $X = +5, +20$ ft. The 20 ft mark is the right end of the beam, and the 5 ft point is where the curve would cross the zero axis if it continued to the left of the 10 ft mark.

SAMPLE PROBLEM 13.3
Beam with Distributed Load

Problem

Given the information in the sketch below, use a shear diagram (a) to determine the location of the maximum moment and (b) to draw a moment diagram.

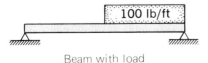

Beam with load

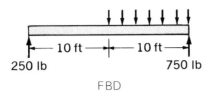

FBD

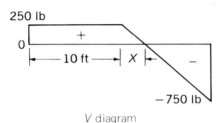

V diagram

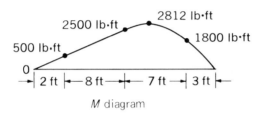

M diagram

Solution (a)

To determine the location of the maximum moment, a simple proportion is set up for the shear diagram. Because of the distributed load, we know that the shear diagram changes at the rate of 100 lb for every foot. Also, by observing the shear diagram, we note that it changes from 0 lb to 250 lb in X ft. Therefore,

$$\frac{X \text{ ft}}{250 \text{ lb}} = \frac{1 \text{ ft}}{100 \text{ lb}}$$

$X = 2.5$ ft

Solution (b)

Cut the shear diagram at the 2 ft mark, and look left. The area of the shear diagram is 250 lb × 2 ft, or 500 lb·ft—the value of the moment at the 2 ft mark. Similarly, we find the area of the shear diagram

at 10 ft mark: $+250 \times 10 = 2500$ lb·ft

at 12.5 ft mark: $+2500 + \frac{1}{2}(250 \times 2.5) = 2812$ lb·ft at 17 ft mark: $+ 2812 - \frac{1}{2}(450 \times 4.5) = 1800$ lb·ft

at 20 ft mark: $+ 2812 - \frac{1}{2}(750 \times 7.5) = 0$ lb·ft

METRIC SAMPLE PROBLEM 13.4
Overhanging Beam with Distributed Load

Problem

A 10 m beam has knife-edge supports placed 2 m from each end and an evenly distributed load totaling 200 kg. Draw the shear and moment diagrams. (200 kg = 1960 N.)

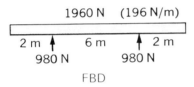
FBD

Solution

Since the general outlines of the diagrams are known, and the moment value at any location equals the shear area to that point, we need check only the locations where the curves change directions.

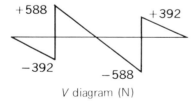
V diagram (N)

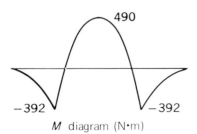
M diagram (N·m)

$V_{2m(-)} = -196(2) = -392$ N
$V_{2m(+)} = -392 + 980 = +588$ N
$V_{8m(-)} = +588 - 196(6) = -588$ N
$V_{8m(+)} = -588 + 980 = +392$ N
$M_{2m} = 1/2[-392(2)] = -392$ N·m
$M_{5m} = -392 + 1/2[588(3)] = 490$ N·m
$M_{8m} = 490 - 1/2[588(3)] = -392$ N·m

Note: You can see that the shear diagram crosses the zero shear axis at the 2 m, 5 m, and 8 m locations, which are the maximum moment locations.

POINT OF INFLECTION

Figure 13.3 shows a loaded beam, the shear diagram, the moment diagram, and also a diagram called the *elastic curve*. The elastic curve is a line sketch of the side view of the beam under load. Stated in more

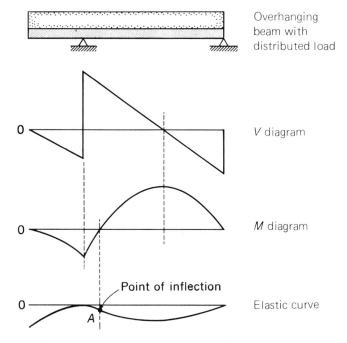

FIGURE 13.3 Shear Diagram, Moment Diagram, Elastic Curve, and Point of Inflection

technical terms, the elastic curve is the edge view of the neutral surface of the beam. For our purposes, the edge view may be thought of as the longitudinal axis. The elastic curve can be demonstrated by observing the deflection of a metre stick simulating a loaded beam. Algebra texts define the *point of inflection* as a point on a curve where the curvature reverses. The point of inflection on a beam is shown as point A on the elastic curve in Figure 13.3. Notice that the point of inflection on the elastic curve is at the same position on the beam as the point where the moment curve crosses the zero moment axis. Therefore, if it becomes necessary to find the point of inflection, we need only find the roots of the moment curve. It is not uncommon for engineers to place a hinge at the point of inflection on bridge structures.

HINGED BEAM

The hinge in a hinged beam cannot transmit a moment but can transmit a force. Generally, the hinge is placed at the point of inflection. You may feel that, in actual practice, moving loads on a bridge would tend to change this inflection point. Nevertheless, the hinge cannot be moved. If loads do change slightly, the reactions at the supports have to adjust for a zero moment at the hinge. Also, note that the beam in Sample Problem 13.5 has three supports and would be statically indeterminate without the hinge. To solve this problem, divide the beam at the hinge, and treat it as two separate beams. You cannot solve the left-hand member until the right-hand member has been solved. By drawing a free body diagram of the right-hand member, we can sum moments about reaction R_3 and find that the reaction of the hinge on the right-hand member is *up* and 100 lb. With this information, we begin to solve the left-hand member. Notice that the free body diagram indicates that the hinge reaction is *down* on the left-hand member. This follows logically: If the left member is holding *up* the right member (which it has to do), the right member must be acting *down* on the left member.

SAMPLE PROBLEM 13.5
Hinged Beam

Problem

Given the loaded hinged beam in the top sketch, (a) draw the FBD and calculate the reactions, (b) construct the shear diagram, and (c) construct the moment diagram.

254 CHAPTER THIRTEEN

Solution

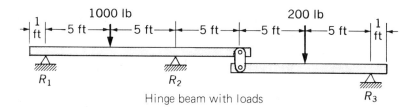

Hinge beam with loads

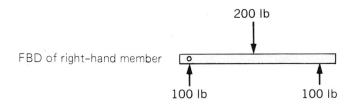

FBD of right-hand member

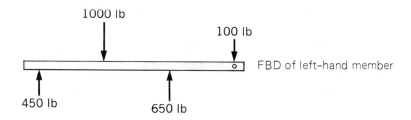

FBD of left-hand member

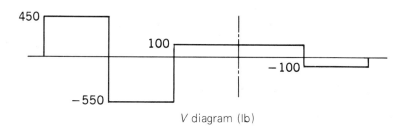

V diagram (lb)

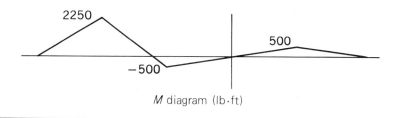

M diagram (lb·ft)

SUMMARY

Internal shear forces and internal resisting couples are part of the statics equilibrium equations:

$$\Sigma F = 0 = +\Sigma_{\text{applied forces}} - \text{internal resisting shear force}$$
$$\Sigma M = 0 = +\Sigma_{\text{applied moments}} - \text{internal resisting couple}$$

The shear diagram is a graphical presentation of the externally applied forces. The moment diagram is a graphical presentation of the externally applied moments.

Shear and moment diagrams are related to each other. The area of the shear diagram up to any selected location on the beam is equal to the moment value at that location.

The point of inflection of the elastic curve of a beam is where the moment curve crosses the zero moment axis.

Hinges may be placed in a beam at the inflection points.

PROBLEMS

Shear Diagrams

For Problems **1–14**, refer to Figures 13.4 through 13.17, and construct the appropriate shear diagrams.

15. (SI Units) Given a simple beam with a span of 10 m and an evenly distributed load of 60 kg/m, construct the shear diagram. Indicate the values where the curves reach the maximums and minimums. Shear forces are to be in newtons, and a simple beam is supported at its ends.

Moment Diagrams

For Problems **16–29**, refer to Figures 13.4 through 13.17, and construct the appropriate moment diagrams.

30. Change the length of the beam in Figure 13.4 to 30 ft. Construct the shear and moment diagrams. Compare these diagrams with

those for Problem 1. Why, in general, can a beam of given cross-sectional shape carry a larger load on a shorter span?

31. (SI Units)   Construct the moment diagram for metric Problem 15.

Moment Equations

For Problems **32–36**, refer to Figures 13.5, 13.14, 13.15, 13.17, and 13.18. Use the shear diagram to determine the location of the maximum moment and to draw the moment diagram. Write the moment equations.

37. (SI Units)   Write the moment equations for metric Sample Problem 13.4.

Inflection Point

38. Refer to Sample Problem 13.2. Locate the inflection point.
39. Refer to Sample Problem 13.5. Locate the point where a second hinge could be placed.
40. Refer to Figure 13.18. Locate the inflection point(s).
41. (SI Units)   Locate the inflection points in Sample Problem 13.4.

Hinged Beams

42. Refer to Sample Problem 13.5. Will the beam support a 500 lb load in place of the 200 lb load?
43. Refer to Figure 13.19. Calculate the support reactions, and draw the shear and moment diagrams.

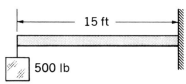

FIGURE 13.4   Problems 1, 16, and 30

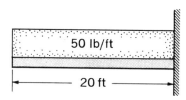

FIGURE 13.5   Problems 2, 17, and 32

Shear and Moment Diagrams 257

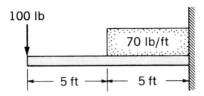

FIGURE 13.6   Problems 3 and 18

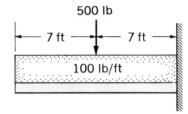

FIGURE 13.7   Problems 4 and 19

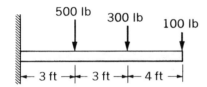

FIGURE 13.8   Problems 5 and 20

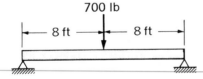

FIGURE 13.9   Problems 6 and 21

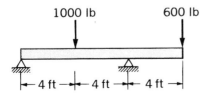

FIGURE 13.10   Problems 7 and 22

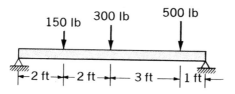

FIGURE 13.11   Problems 8 and 23

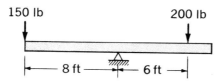

FIGURE 13.12   Problems 9 and 24

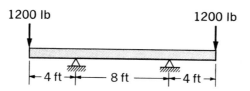

FIGURE 13.13   Problems 10 and 25

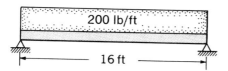

FIGURE 13.14   Problems 11, 26, and 33

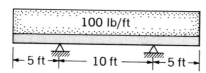

FIGURE 13.15   Problems 12, 27, and 34

Shear and Moment Diagrams 259

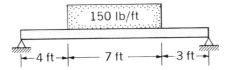

FIGURE 13.16 Problems 13 and 28

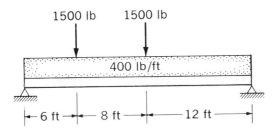

FIGURE 13.17 Problems 14, 29, and 35

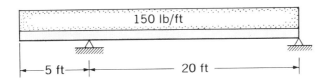

FIGURE 13.18 Problems 36 and 40

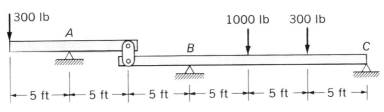

FIGURE 13.19 Problem 43

COMPUTER PROGRAM

Beam reactions and maximum moments can be solved with the following program. The moment values will be at the supports and where the shear diagram crosses the zero axis. Beams can be simply supported or overhanging. Loads must be evenly distributed over the full length of the beam. Units should be in pounds and feet.

CHAPTER THIRTEEN

```
10   REM  MAXMOMENT. THIS PROGRAM WILL FIND THE MOMENTS AT THE REACTIONS A
     ND WHERE THE SHEAR DIAG. CROSSES THE ZERO AXIS-FOR EVENLY DISTRIBUTE
     D LOADS ACROSS THE FULL LENGTH OF THE BEAM.  BEAMS MAY BE SIMPLE SPA
     NS OR OVERHANGING.
20   PRINT "TYPE IN THE TOTAL LOAD, BEAM LENGTH, DISTANCE OF 'A' FROM LEFT
      END, AND DISTANCE OF 'B' FROM LEFT END."
30   INPUT W,L,DA,DB
40   REM  -- FIND REACTIONS FIRST --
50   LET B = (W * ((L / 2) - DA)) / (DB - DA)
60   LET A = W - B
70   PRINT
80   PRINT "REACTION AT A= ";A;" LB."
90   PRINT "REACTION AT B= ";B;" LB."
100  PRINT
110  REM  -- FIND WEIGHT PER FOOT, 'C' --
120  LET C = W / L
130  IF DA = 0 THEN  GOTO 300
140  LET MA =  - (C * DA ^ 2) / 2
150  PRINT "MOMENT AT 'A'= ";MA;" LB-FT."
160  LET MB =  - (C * DB ^ 2) / 2 + (A * (DB - DA))
170  PRINT
180  PRINT "MOMENT AT 'B'= ";MB;" LB-FT."
190  REM  -- FIND MAX MOMENT, 'MM' AND DISTANCE FROM LEFT END, 'X' --
200  LET X = ((A - (C * DA)) / C) + DA
210  LET MM =  - ((C * X ^ 2) / 2) + (A * (X - DA))
220  PRINT
230  PRINT "MAXIMUM MOMENT = ";MM;" LB-FT AT A DISTANCE OF ";X;" FT FROM
     THE LEFT END."
240  END
300  PRINT "MOMENT AT 'A'= 0."
310  GOTO 160
```

14

Bending Stress Formula

OBJECTIVES

This chapter discusses the formula that relates tensile and compressive stresses in a beam to the applied load. The concepts necessary for deriving the formula are presented so that you can have a better understanding of why the formula is used. You will not be expected to derive a formula but to apply it properly. If a specified beam and load are given, you should be able to determine the tensile and compressive stresses (1) for beams with rectangular or circular cross sections and (2) for standard steel beams. Also, given a specified beam and its length, you should be able to determine the allowable concentrated or distributed load.

INTRODUCTION

Galileo (1564–1642) investigated stresses in beams. But Charles Coulomb (1736–1806), the French scientist and inventor who is probably best known for his work in magnetism and electrostatics, was the first to determine the location of the neutral axis in a beam. This axis location is essential for the proper use of the bending stress formula.

EFFECTS OF BENDING STRESSES

In the previous chapter, we indicated that the resisting force system usually acting on any given cross section of a beam consists of the resisting shear force and the resisting couple. In most cases of beam design, the bending stresses produced by the resisting couple are a more serious consideration than the stresses produced by shear forces. This chapter investigates the effect of tensile and compressive bending stresses on the cross section of a beam and their relationship to the applied load.

Bending stresses on a beam cross section vary in intensity. Figure 14.1 illustrates the variation in the bending stresses. As long as the stresses are below the elastic limit for the material being used, the stresses will vary directly as the distance from the neutral axis, axis x-x in Figure 14.1(b). The *neutral axis* is defined as the axis on the cross section where the bending stress is zero. In all of our discussions, the neutral axis coincides with the centroidal X-axis of the cross section. Beams that are curved before loading usually do not have this coincidence and are not discussed in this chapter.

Stack of papers on flat surface

Bottom sheet in tension; top sheets buckling, indicating compression

(*a*) Bending Stresses on a Stack of Papers

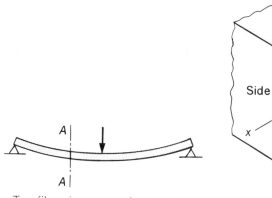

Top fibres in compression; bottom fibres in tension

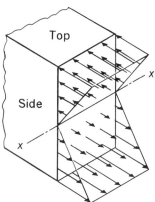

Beam cross section at A-A showing variations in stress

(*b*) Bending Stresses on a Beam

FIGURE 14.1 Tensile and Compressive Stresses Caused by Bending of Beams

THE BENDING STRESS FORMULA

If the beam is symmetrical about the neutral axis of the cross section, the maximum tensile stress will equal the maximum compressive stress. The *bending stress formula* relates maximum stress to the moment value of the resisting couple. In order to understand the theoretical discussion pertaining to the bending stress formula, we must regard the load as tending to bend the beam. This bending causes strains that result in stresses (see Figure 14.2). The cross sections at A and B are parallel to each other before loading and inclined as shown after loading. The distance between the cross sections remains unchanged along the centroidal axis. However, at the top of the beam, the two cross sections are bent toward each other, as indicated by A' and B'. If the deformation is measured and strain determined, then the compressive stress can be determined from the relationship $E = \sigma/\epsilon$. As we go down the cross section, starting at A', it is obvious that the compressive strain decreases until there is no deformation (therefore no stress) at the centroidal X-axis. Continuing toward the bottom of the beam, the tensile strain increases until a maximum is reached at the bottom fibre. In designing beams, it is important to know what the maximum tensile and compressive stresses will be under various types of loads.

The derivation of the bending stress formula makes use of the moment equation $M = F \cdot d$. The basic stress formula $\sigma = F/A$ can be used only for extremely small areas because its use implies that stress

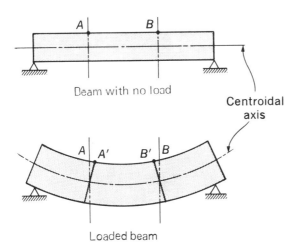

FIGURE 14.2   Illustration of Bending Strains

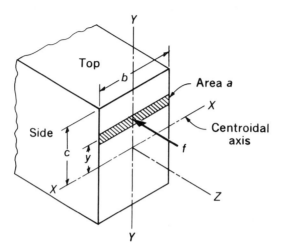

FIGURE 14.3 Cross Section of a Beam under Load and a Force f Acting on a Small Rectangular Area a

does not vary across the corresponding area. In fact, we have to go to an infinitely small area, as designated by a in Figure 14.3. This figure shows a beam cross section and indicates that a force f is applied to the area a at a distance y from the centroidal axis (the neutral axis).

The following discussion describes the derivation of the bending stress formula. Table 14.1 lists the steps. The moment of f (see Figure 14.3) acting on the small area a is designated m and is detailed in Step 1 of Table 14.1. The moment center is on the centroidal X-axis. Keep in mind that the magnitude of f varies, depending on the position of the area on which it is acting. The distance y also varies, depending on the position of the small area; its maximum value is designated c. Step 2 in the table indicates that the total moment value on the whole cross section is the sum of all of the moments acting on all of the small areas in the cross section. The purpose of Step 3 is to get the moment in terms of stress. Since the area a is so small, we may assume that there is no change in force across its face and that $\sigma'(a)$ can be substituted for f. The formula in Step 3 is not entirely satisfactory because it should be in terms of the maximum stress at distance c from the X-axis. Step 4 indicates how the maximum stress is substituted in the formula by using the proportion $\sigma'/y = \sigma_{max}/c$. Rearranging, we have $\sigma' = (\sigma_{max}/c) \cdot y$. Step 5 is a rearrangement of terms. At Step 6 it is recognized that part of the formula represents the moment of inertia of an area (discussed in Chapter 5). I is substituted for $\Sigma y^2(a)$. Step 7 gives us the most common form of the bending stress formula by rearranging terms.

Bending Stress Formula

TABLE 14.1 Derivation of the Bending Stress Formula from $M = F \cdot d$

Step	Equation	Comments
1	$m = f \cdot y$	m is the moment of f acting on area a. The moment center is on the centroidal X-axis.
2	$M = \Sigma_{-c}^{+c} m = \Sigma_{-c}^{+c} f \cdot y$	M is the total moment applied to the whole cross section and is therefore the sum of all the individual moments.
3	$M = \Sigma_{-c}^{+c} [\sigma'(a)](y)$	From the equation $\sigma' = f/a$, substitute $\sigma'(a)$ for f.
4	$M = \Sigma_{-c}^{+c} \left[\dfrac{\sigma_{max}}{c} \cdot y(a) \right](y)$	Since the magnitude of σ' varies directly as y, we have the relation $\sigma'/y = \sigma_{max}/c$. Substitute $(\sigma_{max}/c) \cdot y$ for σ'.
5	$M = \dfrac{\sigma_{max}}{c} \Sigma_{-c}^{+c} y^2(a)$	Remove constants from summation, and collect y terms.
6	$M = \dfrac{\sigma \cdot I}{c}$	Substitute I for $\Sigma_{-c}^{+c} y^2(a)$, and drop the subscript.
7	$\sigma = \dfrac{M \cdot c}{I}$	Rearrange terms for the most common form of the bending stress formula.

Bending stress formula: $\sigma = \dfrac{M \cdot c}{I}$

where
σ = maximum stress appearing on outer fibre of beam, in psi
M = moment of bending forces (couple) on cross section, in lb·in.
c = distance from centroidal X-axis to outer fibre, in inches
I = moment of inertia of cross sectional area, in in.^{4} (inches raised to fourth power)

The following conditions must prevail when you use the bending stress formula:

1. Assuming the beam is horizontal, loads and reactions are applied vertically and along the centroidal Y-axis. No applied force component is allowed along the longitudinal Z-axis of the beam.
2. The cross section of the beam must be symmetrical about the centroidal Y-axis. (Refer to Figure 14.4.)

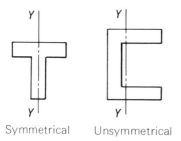

FIGURE 14.4 Beam Cross Section (Symmetrical Condition Must Exist)

3. The stresses involved must not exceed the proportional limit of the material.
4. The beam must be straight before loading.
5. The material of the beam must have the same modulus of elasticity in tension as in compression.
6. The beam must have sufficient lateral support to prevent buckling sidewise. For simplification, this condition is assumed in this text.
7. Cross sections are plane before bending and are assumed plane after bending.

The application of the bending stress formula is described in Sample Problems 14.1 and 14.2 and in metric Sample Problem 14.3.

SAMPLE PROBLEM 14.1
Application of the Bending Stress Formula

Problem

A W 8 × 21 beam supports a load that produces a maximum moment of 25 000 lb·ft. (a) What is the maximum bending stress developed in the beam? (b) At the position of maximum moment, what bending stress is developed 2 in. below the top of the beam?

Solution (a)

The problem describes a wide-flange beam, approximately 8 in. deep and weighing 21 lb/ft. Unless otherwise specified, S-beams and W-beams are placed with their webs vertical for vertical loads. From the Appendix tables, the moment of inertia I about the X-axis is 75.3 in.^{4}; the c-distance is 8.28/2, or 4.14 in. We convert lb·ft to lb·in.: 25 000 lb·ft

× 12 in./ft = 300 000 lb·in. So, the maximum bending stress developed in the beam is:

$$\sigma = \frac{M \cdot c}{I} = \frac{300\,000 \cdot 4.14}{75.3} = 16\,500 \text{ psi}$$

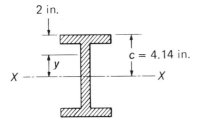

Solution (b)

$$\frac{\sigma'}{y} = \frac{\sigma}{c}$$

$$\frac{\sigma'}{2.00} = \frac{16\,500}{4.14}$$

$$\sigma' = 7970 \text{ psi}$$

SAMPLE PROBLEM 14.2
Application of the Bending Stress Formula

Problem

A hollow timber beam with the cross section as illustrated has an allowable bending stress of 1300 psi. (a) What distributed load can be placed on the beam? (b) What is the top fibre bending stress 2 ft from the left end?

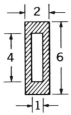

Cross section

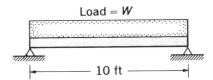

Solution (a)

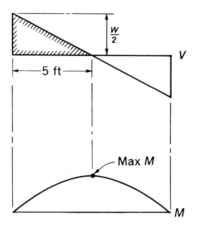

The maximum moment equals the area of the shear diagram up to a 5 ft location:

$$M_5 = \left(\frac{1}{2}\right)\left(\frac{W}{2}\right)(5 \text{ ft} \times 12 \text{ in./ft}) = 15W \text{ lb·in.}$$

In this case, the c-distance is 1/2 the overall height. So,

$c = 3$ in.

$$I = \frac{bh^3}{12} \text{ (for rectangle)}$$

$$I = I_o - I_i = \frac{2(6)^3}{12} - \frac{1(4)^3}{12} = 30.67 \text{ in.}^4$$

where I is the moment of inertia of the net area, I_o is related to the outer rectangle, and I_i is related to the inner rectangle. Then,

$$\sigma = \frac{M \cdot c}{I}$$

$$M = \frac{\sigma \cdot I}{c}$$

$$15W = \frac{\sigma \cdot I}{c}$$

$$W = \frac{\sigma \cdot I}{15 \cdot c} = \frac{1300 \cdot 30.67}{15 \cdot 3} = 887 \text{ lb}$$

Solution (b)

$$M_2 = \frac{W}{2} \times 24 \text{ in.} - (88.7 \text{ lb/ft} \times 2 \text{ ft} \times 12 \text{ in.}) = 8520 \text{ lb·in.}$$

$$\sigma = \frac{M \cdot c}{I} = \frac{8520 \cdot 3}{30.67} = 835 \text{ psi}$$

METRIC SAMPLE PROBLEM 14.3
Application of the Bending Stress Formula

Problem

A simply supported wooden beam 6 m long has a rectangular cross section 60 mm high and 20 mm wide. The allowable bending stress is 13 MPa. What distributed load, in kilograms, can the beam support? Disregard the weight of the beam.

Solution

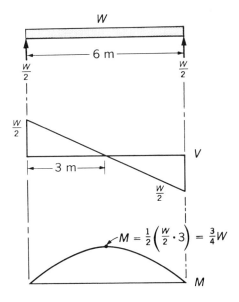

CHAPTER FOURTEEN

$$I = \frac{bh^3}{12} = \frac{0.02(0.06)^3}{12} = 3.6(10)^{-7} \text{ m}^4$$

$$\sigma = \frac{M \cdot c}{I} = \left(\frac{3}{4}\right)\left(\frac{W \cdot c}{I}\right)$$

$$W = \frac{4\sigma \cdot I}{3 \cdot c} = \frac{4 \cdot 13(10)^6 \cdot 3.6(10)^{-7}}{3(0.03)} = 208 \text{ N} = \frac{208}{9.8} = 21 \text{ kg}$$

The moments of inertia for various cross-sectional shapes are presented in the Appendix. This information is repeated in Figure 14.5 for your convenience. In all of the cases shown, the moment of inertia is taken about the X-centroidal axis.

You should study the bending stress formula and determine what happens to the stress when each of the three variables M, c, and I is changed. For instance, if M and c are held constant but I is doubled, what happens to the stress? How can a square beam be redesigned to increase its moment of inertia I without increasing the amount of material used? Would this explain why a 2 × 8 timber is used as a floor joist rather than a 4 × 4, even though both have approximately the same cross-sectional area? Why are standard steel beams flanged on their outer dimensions instead of in the middle, as shown in Figure 14.6?

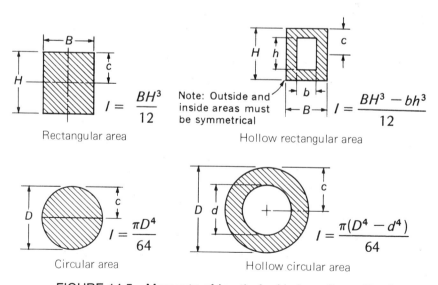

FIGURE 14.5 Moments of Inertia for Various Cross Sections

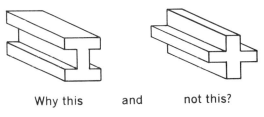

Why this   and   not this?

FIGURE 14.6   Beam Shapes

SUMMARY

The bending stress formula is generally considered the most critical formula in beam design. It relates the maximum tensile and compressive stresses in a beam to the applied load through use of the moment resisting the load.

Bending stress formula: $\sigma = \dfrac{M \cdot c}{I}$

PROBLEMS

Data for standard steel beams and for timbers may be found in the Appendix. Unless otherwise specified, disregard the weight of the beam.

1. An S 3 × 7.5 steel beam, shown in Figure 14.7, supports a maximum moment of 30 000 lb·in. at the wall. What is the bending stress at the outer fibres of the beam (a) at the wall and (b) half-way to the wall?
2. Refer to Problem 1. If the allowable stress in the beam is 22 000 psi, what moment may be applied at the wall?

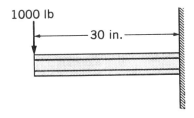

FIGURE 14.7   Problem 1

3. Figure 14.8 shows a loaded cantilever beam. The cross section is also shown. $I = 30.34$ in.^{4}, and the centroidal X-axis is 2.96 in. above the base.
   (a) What bending stress is in the top fibre at the wall?
   (b) What bending stress is in the bottom fibre at the wall?
   (c) Invert the beam and answer (a) and (b).
4. Refer to Problem 3. If the beam is cast iron (stronger in compression than in tension), which way should the beam be placed for the greatest load?
5. A simple beam (supported at its ends) has a 10 ft span and a circular cross section 3 in. in diameter. Determine the maximum bending stress (a) with a center load of 1000 lb and (b) with a 1000 lb distributed load.
6. The beam in Figure 14.9 is made of two 2 × 4s. Consider each

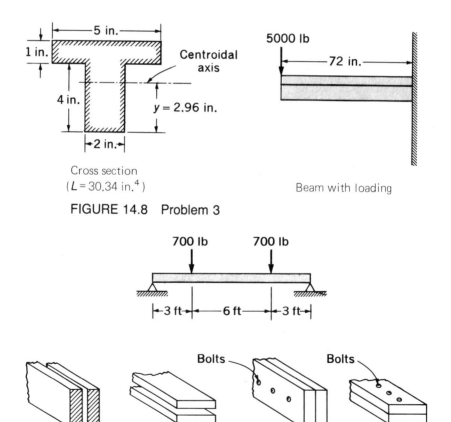

Cross section
($L = 30.34$ in.^{4})

FIGURE 14.8   Problem 3

Beam with loading

FIGURE 14.9   Problem 6

2 × 4 in (a) and (b) to take its fair share of the load. Consider the beam in (c) and (d) to be one solid member.

(a) If the 2 × 4s are placed side by side but not fastened together as in part (a) of Figure 14.9, what maximum bending stress results?

(b) If the 2 × 4s are placed as in part (b) and not fastened together, what maximum bending stress results?

(c) What maximum bending stress results if they are fastened together as in (c)?

(d) What maximum bending stress results if they are fastened together as in (d)?

7. A 6 ft long 2 × 4 timber beam (supported at its ends) has a concentrated load at midspan.

(a) Given an allowable bending stress of 1900 psi and the fact that the cross section is a full 2 in. by 4 in., what load can be placed at midspan?

(b) The actual dressed size of a 2 × 4 is 1½ × 3½. How much would you guess the load would change: 5%, 10%, 20%, 30%, or 50%? Now run out the calculation.

8. The beam in Figure 14.10 is loaded at both ends and has an allowable bending stress of 22 000 psi.

(a) What loads can be placed on both ends of the beam with a cross section as shown in part (a) of Figure 14.10?

(b) Determine the loads with the cross section shown in part (b).

(c) Why should the beam with the cross section in part (b) be stronger when it has the same area and therefore the same amount of material as in part (a)?

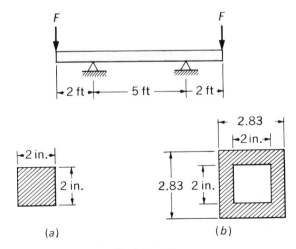

FIGURE 14.10   Problem 8

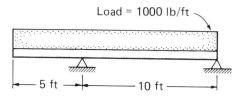

FIGURE 14.11   Problem 10

9. Refer to Problem 8(a). What is the bending stress 1/2 in. below the top fibre and 5 ft from the left end?
10. If the beam in Figure 14.11 is a W 10 × 60 steel beam, what maximum bending stress appears on cross sections 2 ft, 5 ft, and 12 ft from the left end?
11. (Extra Credit) We are given a simply supported beam with a 20 ft span and a 1000 lb concentrated load at midspan.
    (a) Choose a suitable square timber beam to support the concentrated load. The allowable bending stress is 1900 psi. *Hint:* Rearrange the equation $\sigma = (M \cdot c)/I$ to $\sigma = M/(I/c)$ and consider I/c as one term.
    (b) Choose a suitable square steel beam. The allowable bending stress is 22 000 psi.
    (c) Using the average figures given below, which beam is lighter?

    Timber beam weight: 40 lb/ft^3
    Steel beam weight: 490 lb/ft^3

12. (Extra Credit) Refer to Problem 11(b). Use the beam chosen there, but calculate the maximum bending stress developed if the weight of the beam is considered as a distributed load in addition to the 1000 lb concentrated load.
13. (SI Units) A W 8 × 28 cantilever beam 5 m long supports an 800 kg concentrated load at its free end. What is the maximum bending stress in the beam?
14. (SI Units) A simply supported solid round aluminum rod with a span of 1 m and a diameter of 50 mm has a center load of 200 kg. What is the maximum bending stress?
15. (SI Units) Solve Sample Problem 14.3 if the load is concentrated 2 m from the left end.

COMPUTER PROGRAM

The following program will calculate the bending stresses in beams (1) with solid or hollow and rectangular or circular cross sections, and (2) with standard manufactured shapes as listed in the Appendix. *Note:* The following abbreviations are used: OB for outside base, OH for outside height, IB for inside base, and IH for inside height.

```
10   REM   BENDSTRESS
20   PRINT "USE INFORMATION FROM THE MAXMOMENT PROGRAM AS DATA FOR THIS PR
     OGRAM"
21   PRINT "WHICH WILL CALCULATE THE BENDING STRESS."
30   PRINT
40   PRINT "IF THE BEAM IS RECTANGULAR - TYPE 1, IF CIRCULAR - TYPE 2, IF
     A STANDARD MANUFACTURED SHAPE - TYPE 3."
50   INPUT A
60   IF A = 1 THEN   GOTO 100
70   IF A = 2 THEN   GOTO 200
80   IF A = 3 THEN   GOTO 300
90   END
100   PRINT
110   PRINT "TYPE IN ORDER: MAXMOMENT (LB.IN.), OUTSIDE BASE 'B' (IN.), ;O
      UTSIDE HEIGHT 'H' (IN.),"
111   PRINT "INSIDE BASE (TYPE 0 IF CROSS SECTION IS SOLID), INSIDE HEIGHT
      (0 IF SOLID)."
120   INPUT M,OB,OH,IB,IH
130   LET I = ((OB * OH ^ 3) - (IB * IH ^ 3)) / 12
140   GOTO 400
200   PRINT
210   PRINT "TYPE IN ORDER: MAXMOMENT (LB.IN.), OUTSIDE DIAMETER (IN.), IN
      SIDE DIAMETER (IF SOLID TYPE 0)."
220   INPUT M,OD,ID
230   LET I = 3.1416 * ((OD ^ 4) - (ID ^ 4)) / 64
240   LET OH = OD
250   GOTO 400
300   PRINT
310   PRINT "TYPE IN ORDER: MAXMOMENT (LB.IN.), MOMENT OF INERTIA, DEPTH O
      F BEAM CROSS SECTION."
320   INPUT M,I,OH
330   GOTO 400
400   LET S = M * OH / (2 * I)
410   PRINT
420   PRINT "BENDING STRESS = "; INT (S * 10 + .51) / 10;" PSI."
430   PRINT
440   PRINT "MOMENT OF INERTIA = "; INT (I * 100 + .51) / 100;" IN^4."
450   GOTO 90
```

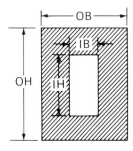

Shear Stress in a Beam

OBJECTIVES

This chapter explains the derivation of the shear stress formula used in beam design and shows how the formula is applied. You will be expected to apply the formula to beams with various cross sections.

INTRODUCTION

Although the bending stresses in beams covered in Chapter 14 are considered to be the most important, shear stresses are important also. To understand shear stress variations from one location to another along a beam, and even the variations over a given cross section, we must first understand horizontal shear stresses. That shear forces, and therefore shear stresses, may exist in a direction along the length of the beam is demonstrated in Figure 15.1. This figure shows a number of common metresticks placed on top of each other and lined up. By pressing down at the center, the metresticks will deflect and slide by each other. This sliding action is caused by horizontal shear forces. If the metresticks were glued together, the glue would resist these shear forces and keep the sticks lined up. Incidentally, if glued together, the metresticks would also not deflect as much under the same load.

There is also a vertical shear force to consider at any given cross section of a beam, as shown in Figure 15.2. The average vertical shear stress on a cross section can be determined by using the standard formula: $\tau = F/A$. Normally, this is not sufficient for design purposes because the shear stress varies too much over the cross section. For example, the maximum shear stress on a beam of rectangular cross section is one and one-half times the average stress value.

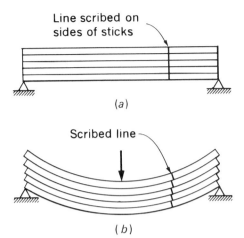

FIGURE 15.1 Demonstration of Horizontal Shear Forces in a Loaded Beam

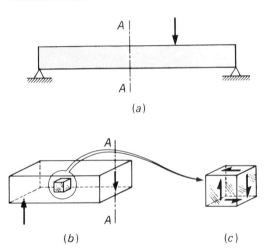

FIGURE 15.2 Relationship of Horizontal and Vertical Shear Forces

Figure 15.2 will aid our understanding of the relationship between horizontal and vertical shear forces. By an examination similar to that made for longitudinal shear stress in a shaft in torsion, it can be concluded that at any point in a beam, the horizontal shear force equals the vertical shear force. To demonstrate this, let's remove an infinitely small cube from the beam in Figure 15.2. If there are vertical shear forces on the cube, then there must be horizontal shear forces, as in Figure 15.2(c), to maintain equilibrium of the cube. These horizontal forces must be equal to each other and to the vertical forces, or

THE BEAM SHEAR STRESS FORMULA

The information following is a discussion of Figure 15.3 and Table 15.1. Derivation of the formula for horizontal shear stress (and therefore vertical shear stress) involves the moment diagram and its relation to the shear diagram. Our object is to find the horizontal shear stress on an internal surface at some specified distance from the centroidal X-axis of a beam cross section. The shear force, which is caused by an imbalance of the bending forces, must be found before finding the

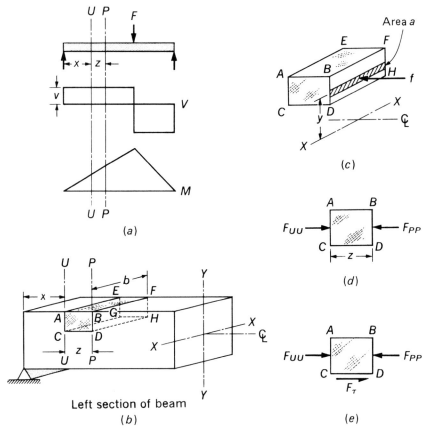

FIGURE 15.3 A Loaded Beam and Details Necessary for Development of Shear Stress Formula

Shear Stress in a Beam

TABLE 15.1 Derivation of the Beam Shear Stress Formula

Step	Equation	Comments
1	$F_{UU} = \sigma_{UU} \cdot A_{ACEG}$ $F_{PP} = \sigma_{PP} \cdot A_{BDFH}$	σ_{UU} and σ_{PP} are average values.
2	$f = \sigma_y(a)$	See Figure 15.3(c); σ_y is the stress distance y from X-axis and f is the force on area a.
3	$F_{UU} = \Sigma f = \Sigma \sigma_y(a)$	Note that this force is the result of bending stresses.
4	$F_{UU} = \Sigma \dfrac{M_{UU} \cdot y(a)}{I}$	Since $\sigma_y = (M \cdot y)/I$ (bending stress formula), substitute for σ_y.
5	$F_{UU} = \dfrac{M_{UU}}{I}\Sigma y(a)$ $= \dfrac{M_{UU}(A \cdot \bar{y})}{I}$	Constants are removed from the summation. Summing up: $\Sigma y(a) = A\bar{y}$.
6	$F_{UU} = \dfrac{M_{UU} \cdot Q}{I}$ $F_{PP} = \dfrac{M_{PP} \cdot Q}{I}$	Let $Q = A\bar{y}$, and substitute.
7	$F_{UU} + F_\tau = F_{PP}$	This is necessary for equilibrium in the horizontal direction.
8	$F_\tau = F_{PP} - F_{UU}$	Rewrite Step 7. Refer to Figure 15.3(e).
9	$F_\tau = \dfrac{M_{PP} \cdot Q}{I} - \dfrac{M_{UU} \cdot Q}{I}$	Substitute $(M \cdot Q)/I$ for F. Refer to Step 6.
10	$F_\tau = \dfrac{Q}{I}(M_{PP} - M_{UU})$	Common factors of terms are removed.
11	$F_\tau = \dfrac{Q}{I}[V(x+z) - V \cdot x]$	From the shear diagram in Figure 15.3(a), we see that $M_{UU} = V \cdot x$, and $M_{PP} = V(x+z)$. Substitute these new values.
12	$F_\tau = \dfrac{Q}{I}(V \cdot z)$	Simplify the equation from Step 11.
13	$\tau \cdot A_{CDGH} = \dfrac{Q}{I}(V \cdot z)$	When stress is constant over an area (such as $CDGH$), $F_\tau = \tau \cdot A$. Substitute for F_τ.
14	$\tau \cdot b \cdot z = \dfrac{Q}{I}(V \cdot z)$	From Figure 15.3(b), $A_{CDGH} = b \cdot z$. Substitute for A.
15	$\tau = \dfrac{V \cdot Q}{I \cdot b}$	Rearrange and simplify.

stress. We choose a block between sections UU and PP of the beam shown in Figure 15.3. This block is labeled $ABCDEFGH$ and has a length b and a measurable height between A and C, but an extremely small width z. The block's bottom area $CDGH$ is the surface on which we want to determine shear stress. The top of the block is chosen to be the top surface of the beam because there is no shear force on this surface. The compressive bending forces on the sides of the block are listed in Step 1 of Table 15.1 and are shown in Figure 15.3(d). We know that the bending forces on areas $BDFH$ and $ACEG$ vary up and down the surfaces on which they are acting, according to their distances from the centroidal X-axis. Therefore, the force f, acting on a small area a, is y in. from the X-axis. See Step 2 in the table and Figure 15.3(c). The force F_{UU} (also F_{PP}) is the total of all the very small forces labeled f. Refer to Steps 3 through 6 of the table.

We must now look at the view of our block showing side $ABCD$ and draw an FBD, as shown in Figure 15.3(d).

Is the block in equilibrium? F_{UU} and F_{PP} are collinear; however, they are not equal because the moment at UU does not equal the moment at PP. See the moment diagram in Figure 15.3(a). Step 6 in the table indicates that the moments must be used to find the forces F_{UU} and F_{PP}. The Qs and Is are equal. Q is the same because the areas are equal and equidistant from the X-axis. I is the same, of course, because it is the I for the whole cross section of the beam. The forces shown in Figure 15.3(d) are *not* in equilibrium because F_{PP} is greater than F_{UU}, and we need an extra force parallel and in the same direction as F_{UU}. See Figure 15.3(e). This extra force F_τ is the shear force on area $CDGH$. We have now placed the block in equilibrium as far as the horizontal forces are concerned. Refer to Step 7 of the table. The formula to find the shear force on the area $CDGH$ is shown developed in Steps 8 through 10, but it is not in convenient form. From Chapter 13, covering shear and moment diagrams, we know that the moment value at any location on the beam is equal to the area of the shear diagram to the left of the location. Therefore, Steps 11 and 12 in the table follow.

The shear stress is assumed to be the same for all points on the surface $CDGH$. Therefore, Steps 13 through 15 follow, and the formula is developed for horizontal (and therefore vertical) shear stress at any level in a beam.

The shear stress formula for beams is as follows:

Beam shear stress formula: $\quad \tau = \dfrac{V \cdot Q}{I \cdot b}$

where
$\quad \tau = $ shear stress at some specified level of beam cross section, in psi

V = vertical shear force applied to beam cross section, taken from shear diagram, in lb
$Q = A \cdot \bar{y}$ for area of cross section above level at which stress is desired, in in.^{3}
$\bar{y}$ = distance from centroidal X-axis of cross section to centroidal X-axis of area above desired stress level, in inches
I = moment of inertia for cross section of beam, in in.^{4}
b = thickness of beam at desired stress level, in inches

The shear stress formula developed in Table 15.1 is used in Sample Problems 15.1 and 15.2 and in metric Sample Problem 15.3.

SAMPLE PROBLEM 15.1
Shear Stress in a Beam

Problem

Given the beam and loading shown in the following sketches, determine the shear stress (a) at the centroidal X-axis and (b) at a level 2 in. below the top surface. The critical section of the beam is that section with the largest value for V. In this case, the critical section is to the right of the load: $V = 900$ lb. *Note:* We do *not* add the shears 900 + 300.

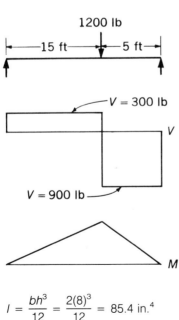

$$I = \frac{bh^3}{12} = \frac{2(8)^3}{12} = 85.4 \text{ in.}^4$$

$b = 2$ in.

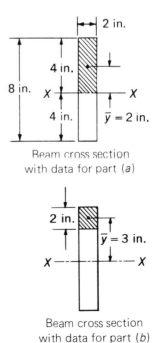

Beam cross section with data for part (a)

Beam cross section with data for part (b)

Solution (a)

Q (for shaded area) $= A \cdot \bar{y} = (2 \cdot 4)(2) = 16$ in.^{3}

$$\tau = \frac{V \cdot Q}{I \cdot b} = \frac{900 \cdot 16}{85.4 \cdot 2} = 84.4 \text{ psi}$$

Solution (b)

Q (for shaded area) $= A \cdot \bar{y} = (2 \cdot 2)(3) = 12$ in.^{3}

$$\tau = \frac{V \cdot Q}{I \cdot b} = \frac{900 \cdot 12}{84.4 \cdot 2} = 63.3 \text{ psi}$$

SAMPLE PROBLEM 15.2
Maximum Shear Stress in a Beam

Problem

Given the information in the following sketches, determine the maximum shear stress (a) in the beam and (b) at a level 1 in. below the top. From the shear diagram, maximum V = 20 800 lb. *Note:* Distributed load is concentrated at its CG in the FBD.

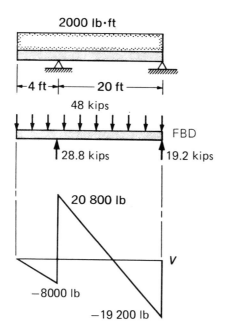
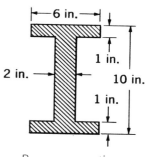

Beam cross section

I (for cross section) = $I_{\text{gross area}} - I_{\text{hollow area}}$

$I_{\text{gross}} = \dfrac{bh^3}{12} = \dfrac{6(10)^3}{12} = 500$ in.^{4}

$I_{\text{hollow}} = \dfrac{bh^3}{12} = \dfrac{4(8)^3}{12} = -171$ in.^{4}

$I = 329$ in.^{4}

Solution (a)

The maximum shear stress is at the centroidal X-axis.

Q (for shaded area) = $\Sigma A\bar{y} = A_1\bar{y}_1 + A_2\bar{y}_2$

$Q = (1 \cdot 6)(4.5) + (2 \cdot 4)(2) = 43$ in.^{3}

$\tau = \dfrac{V \cdot Q}{I \cdot b} = \dfrac{20\,800 \cdot 43}{329 \cdot 2} = 1360$ psi

Solution (b)

$Q = A \cdot \bar{y} = (1 \cdot 6)(4.5) = 27$ in.^{3}

The level 1 in. below the top is a transitional level for stress because b = 2 in. and 6 in. We pick the smaller b because there can be no shear stress on the part of the surface in contact only with the air.

$\tau = \dfrac{V \cdot Q}{I \cdot b} = \dfrac{20\,800 \cdot 27}{329 \cdot 2} = 855$ psi

METRIC SAMPLE PROBLEM 15.3
Maximum Shear Stress in a Beam

Problem

A rectangular cantilever beam 4 m long has a cross section 200 mm high and 70 mm wide. It supports a concentrated end load of 2000 kg. Determine the maximum shear stress in the beam, using the formula

$\tau = \dfrac{V \cdot Q}{I \cdot b}$

where τ is in pascals; V, newtons; Q, metres^3; I, metres^4; and b, metres.

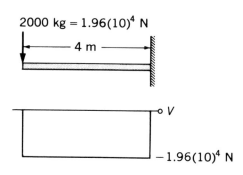
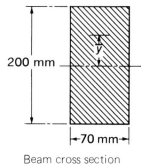

Beam cross section

Solution

$$I = \frac{bh^3}{12} = \frac{0.07(0.2)^3}{12} = 4.7(10)^{-5} \text{ m}^4$$

$$Q = A \cdot \bar{y} = (0.1 \cdot 0.07)(0.05) = 35(10)^{-5} \text{ m}^3$$

$$\tau = \frac{V \cdot Q}{I \cdot b} = \frac{1.96(10)^4 \cdot 35(10)^{-5}}{4.7(10)^{-5} \cdot 0.07} = 21(10)^5 \text{ Pa} = 2.1 \text{ MPa}$$

MANUFACTURED STEEL AND ALUMINUM BEAMS

For manufactured beams with webs (standard S-beams, wide-flange, W-beams, and channels), the shear stress formula $\tau = (V \cdot Q)/(I \cdot b)$ is modified. The formula that may be used is as follows:

$$\tau = \frac{V}{dt}$$

where
V = shear force in pounds
d = total depth of the beam in inches
t = web thickness

This formula is valid for the assumption that the web takes the shear stress and the flanges take the bending stresses, which is not precise, but close enough for practical purposes. Figure 15.4 shows a cross section of a standard beam and the shear stress distribution. Notice that the stress values do not change appreciably over most of the web.

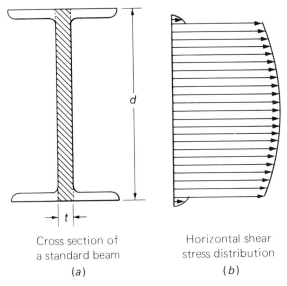

Cross section of
a standard beam
(a)

Horizontal shear
stress distribution
(b)

FIGURE 15.4 Standard Beam and Horizontal Stresses

SHEAR STRESS IN BOLTS IN A BUILT-UP BEAM

When considering a *built-up beam*, as in Figure 15.5, the first step is to determine the level at which the bolts will take a shear stress. The figure shows the level at the bottom edge of the top flange (and the top edge of the bottom flange). If the beam were solid, the web would have to take this stress. Therefore, we use the web thickness for b in the formula

$$\tau = \frac{V \cdot Q}{I \cdot b}$$

and calculate the shear stress at the specified level. We must now calculate the *total force* on this level. Figure 15.5(c) shows the area on which the horizontal shear stress acts (assuming the material is solid and not bolted). This area is the length of the beam in inches multiplied by the thickness of the web. We use the standard formula

$$F = \tau \cdot A$$

This total force is the force the bolts must resist. A separate investigation gives us the allowable force each bolt may take; and from this

CHAPTER FIFTEEN

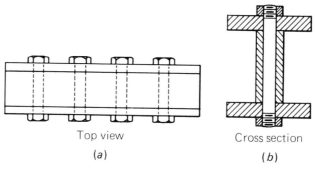

Top view
(a)

Cross section
(b)

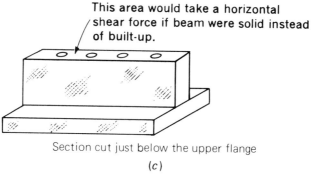

Section cut just below the upper flange
(c)

FIGURE 15.5 A Built-Up Beam

information, we can determine the number of bolts required. For our problems, we will consider the shear force V to be constant along the length of the beam and the bolts to be evenly spaced along the beam. Sample Problem 15.4 illustrates the solution of a built-up beam problem.

SAMPLE PROBLEM 15.4
Built-Up Beam

Problem

The following sketches show the loading on a built-up box beam. The beam is assembled from four 3 in. × 1/2 in. plates and 1/8 in. diameter bolts. The allowable shear stress for the bolts is 15 000 psi. Consider the shank of the bolts in shear. Determine the number of bolts and spacing.

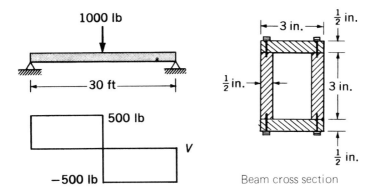

Beam cross section

Solution

From the shear diagram, $V = 500$ lb (constant across the beam).

I (for beam) $= I_{gross} - I_{hollow}$

$I_{gross} = \dfrac{bh^3}{12} = \dfrac{3(4)^3}{12} = 16$ in.^{4}

$I_{hollow} = \dfrac{bh^3}{12} = \dfrac{2(3)^3}{12} = -4.5$ in.^{4}

$I = 11.5$ in.^{4}

Q (for area of top plate) $= A \cdot \bar{y} = (3 \cdot 0.5)(1.75) = 2.62$ in.^{3}

We shall proceed as if the beam were one piece and find the shear stress 1/2 in. below the top. At this level, $b = 1$ in. and 3 in. The 1 in. value is used.

τ (1/2 in. level) $= \dfrac{V \cdot Q}{I \cdot b} = \dfrac{500 \cdot 2.62}{11.5 \cdot 1} = 114$ psi

This value is the shear stress on a horizontal web area 1/2 in. below the top. Therefore, the force on this area (assuming a one-piece beam) is

$F = \tau \cdot A = 114(t)(L) = 114(2 \cdot 1/2 \text{ in.})(30 \cdot 12 \text{ in./ft}) = 41\,000$ lb

This is the force tending to slide the upper plate past the two supporting it. This force must be taken by the bolts.

number of bolts $= \dfrac{41\,000 \text{ lb}}{F_{\text{all per bolt}}} = \dfrac{41\,000}{\tau_{\text{all}} \cdot A_{\text{bolt}}}$

$= \dfrac{41\,000}{15\,000(\pi/4)(1/8)^2} = 223$ bolts

We should specify 224 bolts because there are two rows of bolts on the top of the beam, and each row should have the same number of bolts.

$$\text{spacing} = \frac{\text{total length}}{\text{number of bolts}} = \frac{2(30 \times 12)}{224} = 3.2 \text{ in. between bolts}$$

LIMITATIONS

The assumptions and limitations covering the bending stress formula in Chapter 14 also apply to the shear stress formula. In addition, the beams in this chapter will have the maximum shear stress on the centroidal X-axis of the cross section. There are exceptions. For example, a beam with a triangular cross section has its maximum shear

PHOTO 15.1   A Vishay Stress-Opticon, Used for Photoelastic Stress Analysis

stress halfway between the top and bottom, not along its centroidal axis.

SUMMARY

Shear stresses are zero on the top and bottom fibres of the beam. Shear stresses usually are more critical than bending stresses in short beams.

The shear stress formula for beams is:

$$\tau = \frac{V \cdot Q}{I \cdot b}$$

The shear stress formula for manufactured beams is:

$$\tau = \frac{V}{dt}$$

PROBLEMS

Note: Exclude beam weight unless otherwise specified.

1. Refer to Problems 1, 5, 6, and 8 of Chapter 14, and determine the maximum value for the shear force V in each problem.
2. Determine Q about the designated X-axis for each of the areas shown in Figure 15.6.
3. Given a beam with a cross section 6 in. wide and 14 in. deep and a shear value of $V = 30\,000$ lb, determine and plot the stress for every inch of cross-section height.

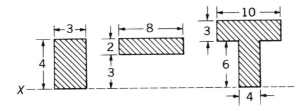

FIGURE 15.6   Problem 2

4. Given a beam with a cross section shown in Figure 15.7 and a shear value of $V = 30\,000$ lb, determine and plot the stress for every inch of cross-section height. (*Note:* The cross-sectional area is the same as for the beam in Problem 3.)
5. Refer to Problems 1, 8, and 10 of Chapter 14. Determine the maximum shear stress developed in each problem.
6. A simply supported 10 ft beam has a 2000 lb concentrated load placed 3 ft from the left end. Determine the maximum shear stress (a) for a standard 4×6 timber beam and (b) for an S 4×7.7 steel beam.
7. A 5 ft cantilever box beam has a 2500 lb force applied to its free end. It has a cross section as shown in Figure 15.8. Determine the maximum shear stress (a) if the force is applied along the X-axis and (b) if the force is applied along the Y-axis.
8. A simply supported beam has a span of 20 ft, a square cross section 6 in. on a side, and a concentrated load of 5700 lb placed X ft from the left end. The maximum allowable shear stress is 150 psi. How close to the left end of the beam can the load be placed?
9. A W 8×35 cantilever beam 15 ft long has an evenly distributed load totaling 7000 lb. Find the maximum shear stress. Include the weight of the beam.
10. A 7 ft long cantilever beam has a 500 lb concentrated load placed at its free end. The cross section is shown in Figure 15.9. The allowable shear stress for the wood dowels is 750 psi. How many dowels are required?
11. (Extra Credit) A simply supported 12 ft beam has a concentrated 1700 lb load at midspan. The cross section is shown in Figure 15.10. Use the AISC code, and determine the number and spacing of the 1/4 in. diameter pins. Pins are A36 steel.
12. (Extra Credit) A built-up beam with a cross section shown in Figure 15.11 is simply supported and has a distributed 200 lb·ft load on its 6 ft span. Use the AISC code, and assume that the maximum shear V applies to the entire length of the beam. The beam is composed of three 1/2 in. $\times$ 3 in. plates bolted together with 1/4 in. bolts. Determine the number and spacing of the bolts. Bolts and plates are A36 steel.
13. (Extra Credit) Starting with $\tau = (V \cdot Q)/(I \cdot b)$, derive an equation for the maximum shear stress in a rectangular beam in terms of the width b and height h. *Hint:* Reduce Q and I to relations of b and h.
14. (Extra Credit) Continue with Problem 13, and show that the shear stress equation for a rectangular beam cross section can be reduced to $\tau = 3V/2A$. (A is the area of the cross section.)

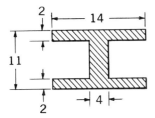

FIGURE 15.7   Problem 4

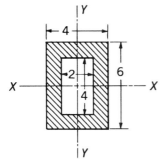

FIGURE 15.8   Problem 7

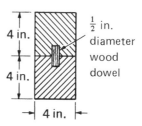

FIGURE 15.9   Problem 10

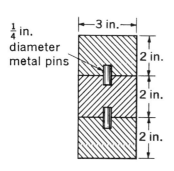

FIGURE 15.10   Problem 11

FIGURE 15.11   Problem 12

15. (SI Units) A 2 × 8 wooden joist, 4 m long, supported at its ends, carries an evenly distributed load of 420 kg. What is the maximum shear developed in the beam? Disregard the weight of the beam.
16. (SI Units) Refer to Problem 14 in Chapter 14, and solve for the maximum shear stress.
17. (SI Units) Calculate the maximum shear stress in metric

Sample Problem 15.3, if the 2000 kg load is evenly distributed over the length of the beam.

18. (SI Units) If the beam in Sample Problem 15.3 were composed of two halves, one on top of the other, how many 3/8 in. bolts are required? Bolts are threaded their entire length and have an allowable shear stress of 14 000 psi.

COMPUTER PROGRAM

The maximum shear stress in a beam can be calculated with the following program. Cross sections must be solid and may be either rectangular or circular or of a manufactured shape listed in the Appendix.

```
10   REM   SHEAR
20   PRINT "THIS PROGRAM WILL CALCULATE THE MAXIMUM SHEAR STRESS IN BEAMS
     WITH CROSS SECTIONS"
21   PRINT "THAT ARE SOLID AND EITHER RECTANGULAR OR CIRCULAR; OR ARE MANU
     FACTURED SHAPES."
30   PRINT
40   PRINT "IF THE CROSS SECTION IS RECTANGULAR - TYPE 1, IF CIRCULAR - TY
     PE 2, AND IF"
41   PRINT "A MANUFACTURED SHAPE - TYPE 3."
50   INPUT X
60   IF X = 1 THEN   GOTO 100
70   IF X = 2 THEN   GOTO 200
80   IF X = 3 THEN   GOTO 300
90   END
100   PRINT
110   PRINT "TYPE IN ORDER: SHEAR LOAD 'V' (LB), HEIGHT OF CROSS SECTION (
      IN.), AND BASE (IN.)"
120    INPUT V,H,B
130    LET S = (3 * V) / (2 * H * B)
135    PRINT
140    PRINT "SHEAR STRESS = "; INT (S * 10 + .51) / 10;" PSI."
150    GOTO 90
200    PRINT
210    PRINT "TYPE IN ORDER: SHEAR LOAD 'V' (LB), AND DIAMETER OF CROSS SEC
       TION (INN.)"
220    INPUT V,D
230    LET S = (V * .424 * 4) / (D ^ 2)
240    GOTO 135
300    PRINT
310    PRINT "TYPE IN ORDER: SHEAR LOAD 'V' (LB), DEPTH OF BEAM (IN.), AND
       WEB THICKNESS (IN.)"
320    INPUT V,D,T
330    LET S = V / (D * T)
340    GOTO 135
```

16

Beam Design

OBJECTIVES

The primary goal of this chapter is to outline the procedure for designing a beam. At this point, you have already applied the proper formulas. This chapter combines that knowledge into a useful pattern. Following the procedures listed here and given a beam design problem, you will be expected to specify the proper manufactured steel or timber beam.

INTRODUCTION

You now have enough information to design beams under certain specified conditions. These conditions are primarily repetitions of the ones given for the bending stress and shear stress formulas:

1. Loads and reactions are applied perpendicular to the longitudinal axis of the beam, so that there is no force component parallel to the longitudinal axis. Also, the loads are applied along the Y-centroidal axis of the beam cross section to avoid twisting the beam.
2. The beam must be uniform over its entire length, and the cross section of the beam is symmetrical about the Y-centroidal axis.
3. The stresses involved must not exceed the proportional limit of the material.
4. The beam is considered straight before loading.
5. The beam must be composed of only one material. This elim-

inates more complicated procedures necessary for reinforced concrete beams and wooden beams with metal strapping.
6. Lateral bracing is not considered. This is frequently determined at a later stage of design.
7. Cross sections of the beam are plane before bending and assumed plane after bending.

Beam design in this chapter is limited to choosing the lightest beam that will do the job, under the assumption that the lighter the beam, the lower the cost. Problems will require you to choose the lightest-weight manufactured steel or timber beam, although the formulas can apply to other construction materials (such as aluminum).

USE OF SECTION MODULUS IN DESIGN

When using the bending stress formula $\sigma = (M \cdot c)/I$ for beam design, both c and I are unknown at first because the beam has not yet been chosen. For this reason, the bending stress formula is rewritten:

$$\sigma = \frac{M}{I/c} = \frac{M}{S}$$

The value I/c is referred to as the *section modulus* and is given the symbol S. In most handbooks, this symbol heads the columns for section modulus in the data tables for steel, aluminum, and timber beam cross sections. Designers rearrange the bending stress formula as follows:

$$S = \frac{M}{\sigma}$$

Design Procedure

Step 1 Construct shear and moment diagrams to determine magnitudes and locations of the maximum moment and maximum shear force. Note that the maximum moment and the maximum shear force are *not* generally at the same location on the beam.

Step 2 Determine or choose the material and its corresponding allowable bending and shear stresses. Materials such as cast iron have different allowable bending stresses for tension than for compression.

Step 3 Apply the bending stress formula $S = M/\sigma$, and determine the size of the beam required from handbook data. Choose the lightest-weight beam that has an S value greater than required.

Step 4 Apply the bending stress formula to determine the required section modulus when the weight of the beam is included in the moment value. If the section modulus of the chosen beam is greater than the required section modulus, go to Step 5. If not, choose the beam with the next-larger section modulus, and repeat this step.

Step 5 Most likely the shear stress is below the allowable. But check anyway, and include the weight of the beam in the calculations. If the shear stress is above the allowable, try the next-larger beam.

Sample Problem 16.1 follows the procedure outlined above.

SAMPLE PROBLEM 16.1
Beam Design

Problem

Using A36 steel, and given the information in the following sketch, determine (a) the lightest-weight steel S-beam, (b) the lightest-weight steel W-beam, and (c) the lightest-weight channel when two are placed back-to-back and bolted together.

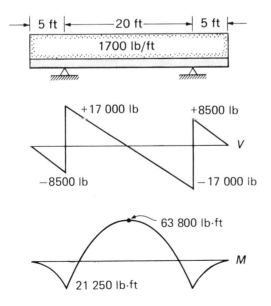

CHAPTER SIXTEEN

Preliminary Solution

Step 1 Take the shear and moment forces, given in the sketch:

$$V_{max} = 17\ 000\ \text{lb}$$
$$M_{max} = 63\ 800\ \text{lb·ft}$$

Step 2 From the Appendix (for A36 steel), take σ and τ values:

$$\sigma_{allowable} = 24\ 000\ \text{psi}$$
$$\tau_{allowable} = 14\ 400\ \text{psi}$$

Step 3 Apply the bending stress formula, section modulus = $S = M/\sigma$:

$$S = \frac{63\ 800 \times 12\ \text{in./ft}}{24\ 000} = 31.9\ \text{in.}^3\ \text{required}$$

Now we are ready to determine the solutions to parts (a), (b), and (c), as stated above.

Solution (a)

To find the lightest S-beam, continue with Step 3 from the preliminary solution above. Choose an S 12 × 31.8 beam ($S = 36.4\ \text{in.}^3$ from the Appendix). The shear diagram will be drawn for the beam's weight only. Then the maximum moment will be calculated from the shear diagram. Both the shear and moment information will be superimposed on the values already obtained for the load.

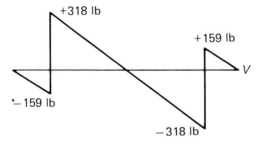

beam weight = 31.8 lb/ft × 30 ft = 954 lb
each support reaction = 477 lb
$M_{max} = -(159 \times 2.5) + (318 \times 5) = 1193\ \text{lb·ft}$

Step 4

$$\text{required } S = \frac{(63\ 800 + 1193) \times 12}{24\ 000} = 32.5\ \text{in.}^3$$

Step 5

$$\text{actual shear stress} = \frac{V}{dt} = \frac{17\,000 + 318}{12 \times 0.35} = 4120 \text{ psi}$$

The actual shear stress is below the allowable, and the required section modulus is below the section modulus of the beam. Therefore, the S 12 × 31.8 beam is appropriate.

Solution (b)

To find the lightest W-beam, continue with Step 3 from the preliminary solution above. Choose a W 10 × 33 beam ($S = 35.0$ in.^{3} from the Appendix).

beam weight = 33 lb/ft × 30 ft = 990 lb

each support reaction = 495 lb

$M_{max} = -(33 \times 15 \times 7.5) + (495 \times 10) = 1238$ lb·ft

Step 4

$$\text{required } S = \frac{(63\,800 + 1238) \times 12}{24\,000} = 32.5 \text{ in.}^3$$

Step 5

$$\text{actual shear stress} = \frac{V}{dt} = \frac{17\,000 + [495 - (33 \times 5)]}{9.75 \times 0.292} = 6300 \text{ psi}$$

The beam chosen, a W 10 × 33, will support the load and its own weight.

Solution (c)

To find the lightest-weight channels placed back-to-back, each channel is assumed to take its fair share of the load, and the required S for each is 32/2, or 16 in.^{3} Continue with Step 3 from the preliminary solution above. Choose a C 10 × 25 beam ($S = 18.2$ in.^{2}).

Cross section

beam weight (for one channel) = 25 lb/ft × 30 ft = 750 lb

each support reaction = 375 lb

M_{max} (one channel) = $-(25 \times 15 \times 7.5) + (375 \times 10) = 938$ lb·ft

Step 4

$$\text{required } S = \frac{[(63\ 800/2) + 938] \times 12}{24\ 000} = 16.4 \text{ in.}^3$$

Step 5

$$\text{actual shear stress} = \frac{(17\ 000/2) + [375 - (25 \times 5)]}{10 \times 0.526} = 1660 \text{ psi}$$

The beam chosen, a C 10 × 25, will support the load and its own weight.

SUMMARY

Beam design is somewhat of a trial-and-error approach because the weight of the beam is not known until a trial solution is made. It is not a good idea to assume a beam weight because you may not end up with the lightest beam. You should follow each step in the procedure.

The section modulus is useful in beam design because it combines the moment of inertia I and the distance from the neutral axis to the outer fibre c into one value: S. This value is given in handbook tables and simplifies the designer's job. S is another symbol for I/c and may be substituted in the bending stress formula: $\sigma = M/S$. For design purposes, the formula is rearranged to:

$$S = \frac{M}{\sigma}$$

PROBLEMS

Note: Where steel beams are specified, use A36 steel and allowable stresses from the AISC code in the Appendix. For wooden beams, use the allowable stresses given for Douglas fir in the Appendix. The dressed sizes of wooden beams are also given in the Appendix.

1. Given that $S = I/c$, derive an equation for the section modulus for a timber beam in terms of b (width) and h (height).
2. Choose the lightest-weight standard S-beam to support the loads in the Sample Problem 13.1 in Chapter 13.
3. Choose the lightest-weight timber beam to support the loading in (a) Figure 13.4, (b) Figure 13.5, and (c) Figure 13.9.

4. Refer to Figure 13.14. Choose the lightest-weight standard S-beam and lightest-weight timber beam. Which beam is lighter?
5. A simple beam (supported at its ends) supports a distributed load of 2000 lb/ft over its entire length of 25 ft.
   (a) Choose the lightest-weight S-beam.
   (b) Choose the lightest-weight W-beam.
6. A 3 ft long cantilever beam supports a total distributed load of 5000 lb. Choose the lightest-weight W-beam.
7. An overhead traveling crane is constructed from a W-shaped steel beam. The span is 40 ft and the maximum load is 2 tons. The load may be moved to a position within 2 ft of either end of the beam. Choose the lightest-weight W-beam. *Hint:* What location of the load will produce the maximum shear force, and what location will produce the maximum bending moment?
8. Figure 16.1 shows a brake lever assembly used to stop a large roller in a paper mill. Specify a rectangular steel lever. Make the depth of the cross section equal to twice the width. Disregard the horizontal friction force.
9. (Extra Credit) Figure 16.2 shows the hallway floor of a school building. The hallway is 12 ft wide. Wooden joists are used as simple beams with a 12 ft span. The spacing between the centers of the joists is 16 in. The loading is 110 lb/ft^2 of floor area. Choose the lightest-weight joist.
10. (Extra Credit) A flat roof is to be placed over the hallway in Figure 16.2. Roof loading is specified at 30 lb/ft^2 (considering

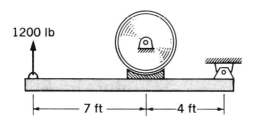

FIGURE 16.1 Problem 8

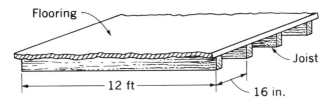

FIGURE 16.2 Problems 9 and 10

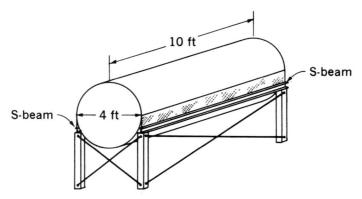

FIGURE 16.3 Problem 11

snow loads and workers making repairs, as well as the weight of the roof itself). If two channels are bolted together back-to-back to form the beams spanning the 12 ft hallway and are spaced 4 ft apart, specify the size of channel to use. Do not consider bolt specifications.

11. (Extra Credit) Figure 16.3 shows a water tank supported by two standard S-beams. The tank and water weigh 10 000 lb. Specify the lightest weight S-beam.

COMPUTER PROGRAM

The following program will aid your selection of a suitable steel beam to support an evenly distributed load and its own weight. Use the program for applicable problems above and to solve this one: Choose the lightest-weight steel beam needed to support a distributed load with a total weight of 40 000 lb and a span of 20 ft. Now, repeat this step and keep the load constant, but shorten the span in increments of 1 ft. Continue until the shear stress is the controlling value. What is the beam and span at this point? (You may want to adjust the program to avoid typing in repetitive data.)

```
10   REM  BEAMDESIGN FOR EVENLY DISTRIBUTED LOAD ON A SIMPLE BEAM FOR STAN
         DARD A-36 STEEL S,W,AND CHANNEL SHAPES.
20   PRINT "BEAM DESIGN.  ALL. BEND. STRESS = 24 000 PSI AND ALL. SHEAR ST
         RESS = 14 400 PSI.  TYPE IN TOTAL LOAD IN POUNDS AND BEAM LENGTH IN
         FEET."
30   INPUT W,L
40   LET S = (W * L * 12) / (8 * 24000)
50   LET I = 1
60   LET Z = 1
```

```
70   PRINT "THE REQUIRED NET SECTION MODULUS = ";S;" CU.IN."
75   PRINT
80   PRINT  TAB( 10);"TRIAL NUMBER ";Z
90   PRINT
100  PRINT "FROM APPENDIX OR HANDBOOK, TYPE BEAM DESIGNATION, DEPTH, WEB
     THICKNESS, SECTION MODULUS, AND WEIGHT PER FOOT."
110  INPUT A$,D,T,SB,C
120  LET S(I) = (12 * L / 8) * (W + C * L) / (24000)
130  IF S(I) > SB THEN  GOTO 300
140  LET SS = W / (2 * D * T)
150  IF SS > 14400 GOTO 400
160  PRINT
170  PRINT "BEAM ";A$;" IS SATISFACTORY.  ITS SECTION MODULUS OF ";SB;" C
     U.IN. IS GREATER THAN THE REQUIRED VALUE OF ";S(I);" AND THE SHEAR S
     TRESS OF ";SS;" PSI IS LESS THAN THE 14 400 PSI LIMIT."
180  END
300  PRINT "YOU MUST CHOOSE A BEAM WITH A LARGER SECTION MODULUS BECAUSE
     THE ACTUAL VALUE OF";S(I);" CU.IN. EXCEEDS THE BEAM'S VALUE."
310  LET Z = Z + 1
320  LET S(I) = S(I) + 1
330  GOTO 80
400  PRINT "YOU MUST CHOOSE A BEAM WITH A LARGER SECTION MODULUS BECAUSE
     THE ACTUAL SHEAR STRESS VALUE OF ";SS;" PSI EXCEEDS THE LIMIT"
410  GOTO 310
```

17

Beam Curvature and Deflection Formulas

OBJECTIVES

This chapter has several objectives. The concepts used to derive beam curvature and deflection formulas are discussed to give you more confidence and a deeper understanding of the formulas. You will be expected to apply the first curvature and the moment-area formulas correctly.

BEAM CURVATURE

For problems dealing with beam curvature, you need two relationships or formulas. One relationship is between the radius of curvature and the maximum stress developed; the other is between the radius of curvature and the moment developed in bending the beam. The following paragraphs discuss these relationships.

The elastic curve of a beam was defined earlier as the neutral axis of a beam. Since all of our beams are straight before loading, the neutral axis and the longitudinal centroidal axis coincide; therefore, we can consider the elastic curve to represent the longitudinal centroidal axis. It is known that the elastic curve assumes the shape of an arc of a circle for whatever distance the moment value remains constant. If couples only are applied to the ends of a beam, the moment for the full length of the beam is constant and the elastic curve for the entire beam assumes the arc of a circle. If the moment value does not remain constant, the elastic curve is assumed to take the shape of the arc of a circle for very short distances.

In Figure 17.1, a very short length l is arbitrarily chosen to separate cross sections AA' and BB'. When the beam is bent, the elastic

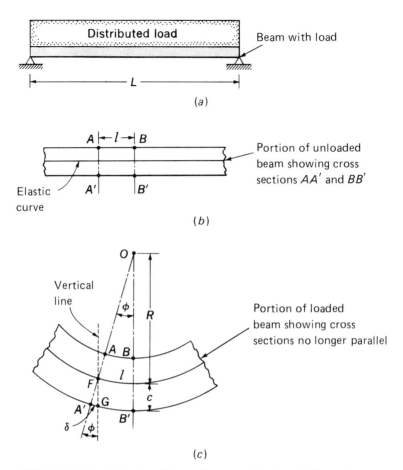

FIGURE 17.1 Relationship of Strain to Radius of Curvature

curve describes an arc of length l with a center at O and a radius R. Turning our attention to the bottom fibres of the beam, we note that the total deformation is labeled δ (*delta*) and that the angle between the cross sections is labeled ϕ (*phi*). Two similar sectors of circles have been formed, $A'OB'$ and $A'FG$. Because two corresponding dimensions of similar sectors are as proportional to each other as any other two corresponding dimensions, we get the relationship

$$\frac{c}{R} = \frac{\delta}{l}$$

This relationship is developed into two beam curvature formulas in Table 17.1. The formulas are:

$$\text{Curvature formula 1:} \quad \frac{c}{R} = \frac{\sigma}{E}$$

$$\text{Curvature formula 2:} \quad \frac{1}{R} = \frac{M}{E \cdot I}$$

where

c = distance from longitudinal axis (elastic curve) to outer fibre in inches
R = radius of curvature measured to elastic curve in inches
σ = tensile or compressive stress in psi
E = modulus of elasticity in psi
M = applied moment in lb·in.
I = moment of inertia of cross section, taken about the centroidal axis, measured in in.[4]

It is assumed that the cross section remains constant for the length of the beam under consideration.

Curvature Formula 1 is used for problems related to coiling material without exceeding the allowable stress. The procedure for

TABLE 17.1 Development of Formulas Relating Radius of Curvature to Bending Stress and Bending Moment (Refer to Figure 17.1)

Step	Formula	Comments
1	$\dfrac{c}{R} = \dfrac{\delta}{l}$	Geometric proportion.
2	$\dfrac{c}{R} = \dfrac{\epsilon \cdot l}{l}$	From Chapter 6, $\delta = \epsilon \cdot l$. Substitute for δ.
3	$\dfrac{c}{R} = \dfrac{\sigma}{E}$	From Chapter 6, $\epsilon = \sigma/E$. Substitute for ϵ.
4	$\dfrac{c}{R} = \dfrac{M \cdot c}{E \cdot I}$	The bending stress formula is $\sigma = (M \cdot c)/I$. Substitute for σ.
5	$\dfrac{1}{R} = \dfrac{M}{E \cdot I}$	The c's cancel out. This is the accepted form for further development of deflection formulas: $c/R = \sigma/E$ (Formula 1, as in Step 3) and $1/R = M/(E \cdot I)$ (Formula 2, as in Step 5).

using this formula is fairly straightforward. Three of the values must be given in the problem or must be obtained from handbook data in the Appendix. The formula is then used to solve for the fourth variable. Sample Problem 17.1 and Problems 1 through 6 at the end of this chapter make use of Formula 1.

SAMPLE PROBLEM 17.1
Curvature of a Band Saw Blade

Problem

A steel band saw blade is 1/32 in. thick by 1/4 in. wide. The bending stress is not to exceed 30 000 psi. What should the pulley diameter be?

Solution

$$\frac{c}{R} = \frac{\sigma}{E}$$

$$R = \frac{c \cdot E}{\sigma} = \frac{(1/32 \cdot 1/2) \cdot 30(10)^6}{30(10)^3} = 15.6 \text{ in.}$$

diameter = 31.2 in.

Usually, a standard-size pulley is chosen with a diameter equal to, or greater than, the answer.

With reference to Formula 1, note that for a given material, σ/E will be a constant. Therefore, if we want to reduce the radius of curvature R, then c must be reduced. This means that for sharper bends, we would need a thinner band saw blade in Sample Problem 17.1. Does this formula indicate to you the reason for using stranded wire rope for ski tows and elevators and stranded electrical conductors in applications subject to frequent bending?

Consider Formula 2 and the familiar situation of walking across a plank supported at its ends. A heavy person causes a larger deflection than a light person; the result is a smaller radius of curvature. The heavier weight means a larger moment M, and the formula indicates that as M increases, R decreases. Curvature Formula 2 will be used later to help develop the deflection formulas.

BEAM DEFLECTION

Frequently, the engineer or technician must design a beam to meet certain deflection limits rather than stress limits. For example, a joist supporting a plaster ceiling must not deflect more than 1/360 of the span length, or the plaster may crack. Beams supporting a turbine-generator set frequently have a deflection limit of 1/2000 of the span. This reduces alignment problems between the generator and turbine shafts. Also, certain statically indeterminate beam problems require deflection formulas to help solve for reactions.

There are a number of approaches for determining beam deflections, but this text will discuss only the *moment-area method*. Generally speaking, beam deflections due to shear forces are negligible; therefore, the moment-area formulas are related to the bending stress formula $\sigma = (M \cdot c)/I$.

Two moment-area formulas are involved. The first provides the angle between the tangents to any two points on the elastic curve. The second gives the vertical distance between a point on the elastic curve and the tangent to any other point on the elastic curve. In most of our problems, we will want this vertical distance between the point and the tangent to be the deflection of the beam; therefore, one reference point on the elastic curve must have a horizontal tangent. Figure 17.2 pictures a center-loaded beam and the information to be obtained from the deflection formulas.

First Moment-Area Formula

Figure 17.2(*b*) shows the angle θ (*theta*) between any two tangents to the elastic curve, which can be determined with the formula:

$$\text{First moment-area formula:} \quad \theta = \frac{A_m}{E \cdot I}$$

where

θ = angle in radians between tangents to elastic curve
A_m = area of moment diagram in lb·in.^{2} between the two positions on elastic curve being considered
E = modulus of elasticity in psi
I = moment of inertia of cross section of beam in in.^{4}

The moment of inertia is taken about the *X*-centroidal axis of the cross section.

This formula states that the angle between tangents to any two points (such as *A* and *B* in Figure 17.2) on the elastic curve is

Beam Curvature and Deflection Formulas   307

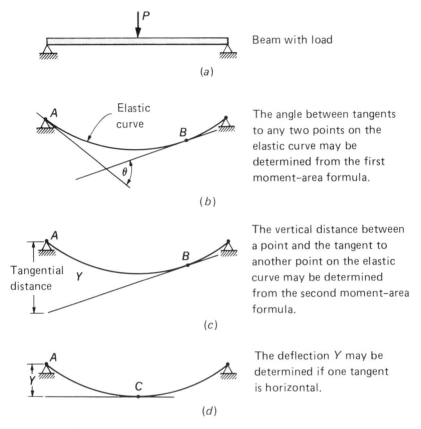

FIGURE 17.2 Illustrations of Information Provided by Moment-Area Formulas

equal to the area of the corresponding portion of the bending moment diagram, divided by $E \cdot I$.

Figure 17.3 shows the development of the first moment-area formula. It is interesting to note that the formula relies on the approximation that an arc length equals a subtending straight line. You may feel this could lead to large errors, but in practical beam applications, angles ϕ and θ are very small, and any resulting errors are negligible.

A beam is chosen with a loading consisting only of couples placed at the beam ends. This means that there are no applied shear forces and that the moment is constant over the full length of the beam (refer to Figure 17.3).

308 CHAPTER SEVENTEEN

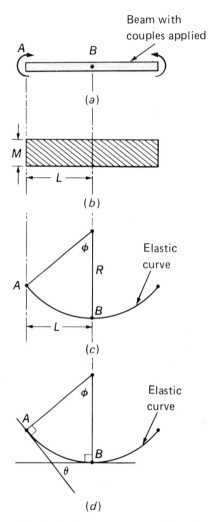

FIGURE 17.3 Concepts Used to Develop First Moment-Area Formula

Procedure for Deriving the First Moment-Area Formula

Step 1 From geometry, arc length = central angle in radians times the radius; therefore, we have

$$AB = \phi \cdot R$$
$$\phi = \frac{AB}{R}$$

Step 2 Let us say that arc length AB equals L in Figure 17.3(c). This is very close to being true when considering the small deflections of beams. Therefore, since $L = AB$, we can substitute L:

$$\phi = \frac{1}{R}(L)$$

Step 3 From Formula 2 of Table 17.1, we have

$$\frac{1}{R} = \frac{M}{E \cdot I}$$

Substituting $M/(E \cdot I)$ for $1/R$ in Step 2, we get

$$\phi = \frac{M(L)}{E \cdot I}$$

Step 4 From Figure 17.3(b), we see that $M \times L$ is the area of the moment diagram between points A and B. Also, from Figure 17.3(d), $\theta = \phi$ because, from geometry, two angles whose sides are perpendicular are either equal or supplementary. By inspection, they are equal. Therefore, we can write

$$\theta = \frac{A_m}{E \cdot I}$$

This formula is used to obtain the second moment-area formula and in deflection problems similar to the type explained in Sample Problem 17.5.

Second Moment-Area Formula

Figure 17.2(c) and (d) show the deflection Y that can be obtained with the second moment-area formula. Figure 17.4 is an enlargement of the moment diagram and the elastic curve shown previously in Figure 17.3. The concepts used to develop the second moment-area formula are detailed below; the measurements are indicated in Figure 17.4.

Procedure for Deriving the Second Moment-Area Formula

Step 1 From geometry, arc length = central angle times the radius; therefore, we have

$$AE = \theta \cdot \bar{x}$$

where $\bar{x}$ is the radius.

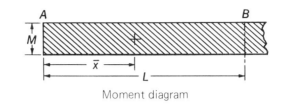

Moment diagram

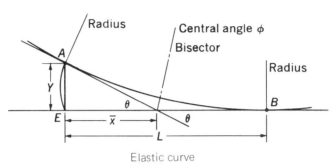

Elastic curve

FIGURE 17.4 Concepts Used to Develop Second Moment-Area Formula

Step 2 Because θ is very small, the vertical distance Y is essentially the same as the arc length; therefore, substitute Y for arc length AE.

$$Y = \theta \cdot \bar{x}$$

Step 3 From the first moment-area formula, we have $\theta = A_m/(E \cdot I)$; therefore, substitute $A_m/(E \cdot I)$ for θ in Step 2.

Second moment-area formula: $\quad Y = \dfrac{A_m \cdot \bar{x}}{E \cdot I}$

Note, that $\bar{x}$ is not only the distance from the vertical at A on the elastic curve to the vertex of θ, but it is also the distance to the centroid of the area of the moment diagram between A and B. Because the vertex of θ falls on the bisector of central angle φ (*phi*) and arc length AB, and since we consider the arc length equal to L, then the vertex of θ (*theta*) is also on the bisector of L. The distance $\bar{x}$ is measured in inches. Also, the values of E and I are considered constant for the full length of the beam.

The deflection Y is positive or negative, depending on the sign of the area of the moment diagram. This is because E and I are positive, and $\bar{x}$ is always taken as being positive. A positive sign for Y indicates that a point on the elastic curve has been deflected up from the tangent to the second point on the elastic curve. An important decision is to determine from which point the centroidal distance $\bar{x}$ should be measured. As an example, the center-loaded beam in Figure 17.2 has its maximum deflection at the midpoint C. Because this point has a horizontal tangent, we consider that point A has been deflected up above C, and $\bar{x}$ is measured from point A. The rule is: *$\bar{x}$ is measured from the deflected point to the centroid of the moment diagram.*

An examination of the second moment-area formula will indicate that the deflection increases if the area of the moment diagram or the moment arm $\bar{x}$ is increased. The deflection decreases if a stiffer material is used (E is increased) or if I is increased.

In the following sample problems, and in most of the problems at the end of this chapter, the weight of the beam is neglected in order to simplify explanations and solutions. Although the second moment-area formula is a general formula that can be used on any beam and loading in this text, you may elect to use specific deflection formulas for specific beam loadings. These specific formulas are available in mechanical and civil engineering handbooks. Some of the extra credit problems include the combination of the applied loads and the beam weight. In such cases, you have the choice of drawing the complete moment diagram and solving for Y directly, or you may solve for the deflection due to each type of loading separately and then add the deflections. One must be careful in this approach because the individual points of maximum deflection for each load may not coincide. The method of adding deflections is called *superposition*.

Procedure for Solving Beam Deflection Problems

Step 1 Draw the shear and moment diagrams.

Step 2 Determine (by inspection, if possible) the location of the point of maximum deflection and the point that has a horizontal tangent. If the location of the point on the elastic curve having a horizontal tangent cannot be determined by inspection, follow the procedure demonstrated by Sample Problem 17.5. The first moment-area formula is needed.

Step 3 Determine the area of the moment diagram between the two locations specified in Step 2.

Step 4 Calculate the $\bar{x}$ (*X-bar*) distance between the centroid of the corresponding area of the moment diagram and the point considered to be deflected from the horizontal tangent to the elastic curve.

312 CHAPTER SEVENTEEN

Step 5 Select the correct modulus of elasticity E, and select or determine the moment of inertia I for the cross section of the beam.

Step 6 Solve for the deflection Y.

Sample Problems 17.2 through 17.6 indicate how to handle cantilever beams with concentrated and distributed loads, overhanging beams with concentrated loads, and a simple beam with an off-center concentrated load.

SAMPLE PROBLEM 17.2
Deflection of a Cantilever Beam

Problem

An S 4 × 9.5 steel beam supports the load shown in the following sketch. What is the deflection (a) under the load and (b) at the end of the beam?

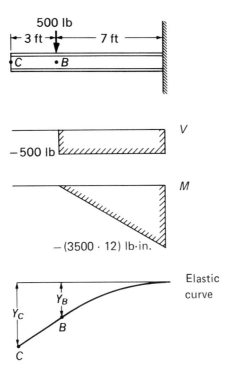

Solution (a)

E for steel and I for the beam are obtained from the Appendix.

$E = 30(10)^6$ psi

$I = 6.79$ in.^{4}

The area of the moment diagram between the wall and point B are then calculated.

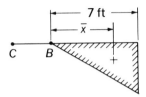

$A = -3500 \cdot 12(7 \cdot 12)\left(\dfrac{1}{2}\right) = -1\,763\,000$ lb·in.^{2}

$\bar{x} = \dfrac{2}{3} \cdot 7 \cdot 12 = 56$ in.

$Y_B = \dfrac{A \cdot \bar{x}}{E \cdot I} = \dfrac{-1.763(10)^6 \cdot 56}{30(10)^6 \cdot 6.79} = -0.485$ in.

Solution (b)

The area of the moment diagram between point C and a point where the tangent is horizontal (the wall) is the same as in part (a).

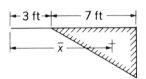

$A = -1.763(10)^6$ lb·in.^{2}

$\bar{x} = 12\left(3 + \dfrac{2}{3} \cdot 7\right) = 92$ in.

$Y_C = \dfrac{A \cdot \bar{x}}{E \cdot I} = \dfrac{-1.173(10)^6 \cdot 92}{30(10)^6 \cdot 6.79} = -0.798$ in.

Comment: The area of the M diagram must be taken between the point where deflection is desired and the point whose tangent is horizontal. The distance $\bar{x}$ is taken from the location of the deflected point to the centroid of the area.

SAMPLE PROBLEM 17.3
Deflection of a Cantilever Beam with More than One Load

Problem

A 4 × 8 timber beam is placed with the 8 in. dimension vertical and supports the loads shown. Determine the maximum deflection of the beam.

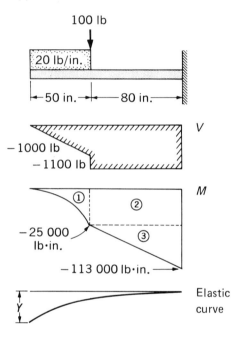

Solution

From the Appendix tables,

$E = 1.76(10)^6$ psi
$I = 111$ in.^{4}

Calculate the area of the moment diagram between the free end of the beam and the wall.

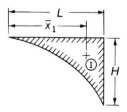

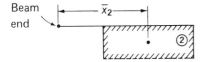

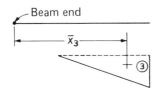

$A \cdot \bar{x} = \Sigma a\bar{x} = a_1\bar{x}_1 + a_2\bar{x}_2 + a_3\bar{x}_3$

$\bar{x}_1 = \dfrac{3}{4}L = \dfrac{3}{4}(50 \text{ in.}) = 37.5 \text{ in.}$

$\text{area} = \dfrac{1}{3}(L \cdot H) = \dfrac{1}{3}(50 \cdot -25\,000)$

$a_1 = 41.6(10)^4 \text{ lb·in.}^2$

$a_1\bar{x}_1 = 37.5 \cdot -41.6(10)^4 = -15.6(10)^6 \text{ lb·in.}^3$

$a_2\bar{x}_2 = -(80 \cdot 25\,000)(90) = -180(10)^6 \text{ lb·in.}^3$

$a_3\bar{x}_3 = -\dfrac{1}{2}(80 \cdot 88\,000)(50 + \dfrac{2}{3} \cdot 80) = -364(10)^6 \text{ lb·in.}^3$

$Y = \dfrac{A \cdot \bar{x}}{E \cdot I} = \dfrac{-(15.6 + 180 + 364)(10)^6}{1.76(10)^6 \cdot 111} = -2.93 \text{ in.}$

Comment: The area of the moment diagram is broken down into parts for easy handling. As in Sample Problem 17.2, the area between the free end of the beam and the wall is important.

SAMPLE PROBLEM 17.4
Deflection of an Overhanging Beam with End Loads

Problem

An S 6 × 12.5 steel beam is loaded as shown in the following sketch. Determine the maximum deflection.

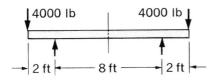

316 CHAPTER SEVENTEEN

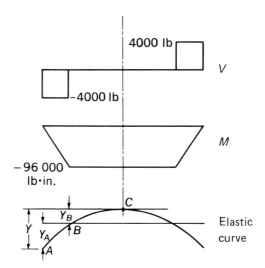

Solution

Because of symmetrical loading, we can determine by inspection that the maximum deflection is at the ends of the beam or in the middle of the beam. Our procedure will be to determine the distances of points A and B from the (horizontal) tangent of point C on the elastic curve. From this information, the deflections of points A and C from their original positions can be determined. Start with the formula $Y = (A \cdot \bar{x})/(E \cdot I)$. From the Appendix tables,

$E = 30(10)^6$ psi
$I = 22.1$ in.^{4}

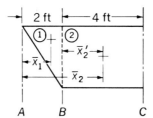

Left section of M diagram

Solve for the maximum deflection.

$$A \cdot \bar{x} = \Sigma A\bar{x} = -(A_1\bar{x}_1 + A_2\bar{x}_2)$$
$$= \frac{1}{2}\left(24 \cdot 96\,000\right)\left(\frac{2}{3} \cdot 24\right) + (48 \cdot 96\,000)(48)$$

$$= 18.41(10)^6 + 221(10)^6 = -239(10)^6 \text{ in.}^3$$

$$Y = \frac{-239(10)^6}{30(10)^6 \cdot 22.1} = -0.360 \text{ in.}$$

$$Y_B = \frac{A \cdot \bar{x}}{E \cdot I}$$

$$A \cdot \bar{x} = A_2 \bar{x}_2' = -48 \cdot 96\,000 \cdot 24 = -110.5(10)^6 \text{ in.}^3$$

$$Y_B = \frac{-110.5(10)^6}{30(10)^6 \cdot 22.1} = -0.167 \text{ in.}$$

$$Y_A = -0.360 - (-0.167) = -0.193 \text{ in.}$$

SAMPLE PROBLEM 17.5
Deflection of a Simple Beam with an Eccentric Load

Problem

A W 8 × 24 steel beam supports the load in the following sketch. Determine the maximum deflection and its location on the beam.

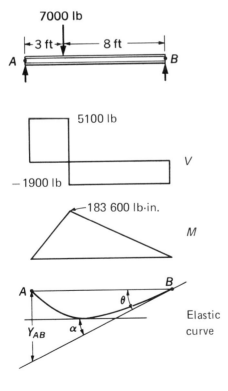

Solution

In this instance, loading is *not* symmetrical. Therefore, the maximum deflection point is not at the center of the beam. Our approach will be to determine the distance Y_{AB} from A to the line tangent to the elastic curve at B. With this information, and knowing distance AB, angle θ can be determined. Angle θ is equal to angle α (*alpha*), the angle between the tangent at B and the tangent to the point of maximum deflection.

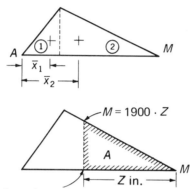

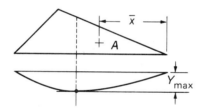

Location of position
of maximum deflection

The first formula states that the angle between tangents is equal to $A_m/(E \cdot I)$ and is used to determine the position of the point of maximum deflection and the maximum deflection. Start with formula $Y_{AB} = (A \cdot \bar{x})/(E \cdot I)$. From the Appendix tables,

$E = 30(10)^6$ psi

$I = 82.8$ in.^{4}

Now, determine angle θ.

$$A \cdot \bar{x} = A_1\bar{x}_1 + A_2\bar{x}_2$$
$$= \frac{1}{2}\left(36 \cdot 183\,600\right)\left(\frac{2}{3} \cdot 36\right) + \frac{1}{2}\left(96 \cdot 183\,600\right)\left(36 + \frac{1}{3} \cdot 96\right)$$
$$= 79.4(10)^6 + 600(10)^6 = 679(10)^6$$

$$Y_{AB} = \frac{679(10)^6}{30(10)^6 \cdot 82.8} = 0.273 \text{ in.}$$

$$\tan \theta = \frac{Y_{AB}}{AB} = \frac{0.273 \text{ in.}}{132 \text{ in.}} = 0.002\ 07 = \theta \text{ (in radians)}$$

$$\alpha \text{ (in radians)} = 0.002\ 07 = \frac{A}{E \cdot I}$$

$$A = \frac{1}{2}(M \cdot Z) = \frac{1}{2}(1900 \cdot Z)(Z) = 950 \cdot Z^2 = 0.002\ 07 \cdot E \cdot I$$

$$Z = \sqrt{\frac{0.002\ 07(30)(10)^6 \cdot 82.8}{950}} = \sqrt{5400} = 73.6 \text{ in.}$$

We now proceed to solve for the maximum deflection.

$$Y_{max} = \frac{A \cdot \bar{x}}{E \cdot I} = \frac{1/2(1900)(73.6)^2\ (2/3 \cdot 73.6)}{30(10)^6 \cdot 82.8}$$

$$= \frac{252(10)^6}{30(10)^6 \cdot 82.8} = 0.101 \text{ in.}$$

Comment: Refer to angle θ; for angles smaller than about 3°, the tangent function is equal to the radian value. When determining the distance Z above, the moment value (1900 · Z) and the area of the moment diagram are taken to the right side because we can solve a right triangle easily.

METRIC SAMPLE PROBLEM 17.6
Deflection of a Cantilever Beam with an End Load

Problem

A Douglas fir cantilever beam 4 m long has a cross section 150 mm high and 40 mm wide and an end load of 40 kg. What is the deflection in mm? Disregard the weight of the beam. The required units in the formula $Y = (A_m \cdot \bar{x})/(E \cdot I)$ are as follows: Y is expressed in metres; A_m, in newton·metres²; $\bar{x}$, in metres; E, in newton/metres² (pascal); and I, in metres⁴.

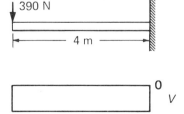

CHAPTER SEVENTEEN

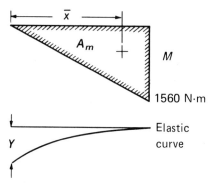

Solution

$$V = 40 \text{ kg} \times 9.8 = 390 \text{ N}$$

$$A_m = \frac{1}{2}(1560 \cdot 4) = 3120 \text{ N·m}^2$$

$$\bar{x} = \frac{2}{3} \cdot 4 = 2.7 \text{ m}$$

$$E = 12(10)^9 \text{ Pa}$$

$$I = \frac{bh^3}{12} = \frac{0.04(0.15)^3}{12} = 1.1(10)^{-5} \text{ m}^4$$

$$Y = \frac{A_m \cdot \bar{x}}{E \cdot I} = \frac{-3120(2.7)}{12(10)^9 \, (1.1)(10)^{-5}} = -637(10)^{-4} \text{ m}$$

$$Y = -64 \text{ mm (rounded off)}$$

SUMMARY

Formula 1 for beam curvature is useful for coiling problems and bending metal bands or wire rope around pulleys:

$$\text{Curvature formula 1:} \quad \frac{c}{R} = \frac{\sigma}{E}$$

Engineers and technicians must be able to determine the deflection of beams under various types of loads. One method of accomplishing this is with the moment-area formulas.

The first moment-area formula provides the angle between two tangents to the elastic curve of the beam:

First moment-area formula: $\quad \theta = \dfrac{A_m}{E \cdot I}$

The second moment-area formula is a very useful general formula for deflection of beams. Remember that $\bar{x}$ is measured from the point of deflection to the centroid of the area of the moment diagram. This statement can cause confusion because a simple beam is considered to have the parts of the beam over the reactions deflected up above the midpoint of the beam. The tangent to the midpoint of a symmetrically loaded simple beam is horizontal, and deflections must be measured to this tangent.

Second moment-area formula: $\quad Y = \dfrac{A_m \cdot \bar{x}}{E \cdot I}$

PROBLEMS

Curvature

1. A steel surveyor's tape is 0.30 in. wide and 0.020 in. thick. It is to be coiled for carrying. What is the smallest diameter coil possible without exceeding a 40 000 psi allowable bending stress?
2. Sheet steel is to be coiled in rolls 3 ft in diameter. If the stress is not to exceed 20 000 psi, what should the thickness of the steel be?
3. A portable horizontal band saw has 8 in. diameter pulleys. What should the blade thickness be if the allowable bending stress is 30 000 psi?
4. A steel railroad rail, shown in Figure 17.5, has a maximum c distance of 2.5 in. about the vertical centroidal axis. The maximum c distance is 2.5 in.

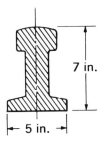

FIGURE 17.5 Problem 4

(a) What is the radius of curvature if the bending stress is not to exceed 15 000 psi?

(b) For high-speed train operation, the radius of curvature can be no smaller than 800 ft. What bending stress is developed?

5. A gardner wants to form a 20 ft diameter circle by bending wooden planks with dimensions 10 ft by 1 ft by 3/8 in. Will the allowable bending stress for wood be exceeded?

6. A 3/8 in. diameter wire rope is used on a ski lift. The diameter of each strand of wire is taken as 1/9 of rope diameter. What diameter pulley will cause a bending stress of 20 000 psi (a) if steel wire is used and (b) if aluminum wire is used?

7. (SI Units) What is the smallest diameter coil possible (in mm) for sheet aluminum 2 mm thick if the allowable bending stress is 100 MPa?

Deflection

In the following problems, neglect the weight of the beam unless otherwise specified, and assume that the elastic limit is not exceeded.

8. A 6 ft long cantilever beam has a concentrated load of 500 lb at its free end. The beam has a square cross section 6 in. × 6 in. What is the maximum deflection (a) for a timber beam, (b) for a steel beam, and (c) for an aluminum beam?

9. Refer to Figure 17.6. The solenoid contact bar is made of copper. What force is required to make contact?

10. A pure moment only of 60 lb·in. is applied to the free end of an aluminum cantilever beam 25 in. long, 2.0 in. wide, and 0.25 in. high. What is the maximum deflection?

11. An S 4 × 9.5 steel cantilever beam 6 ft long supports a 2500 lb concentrated load at its free end and another concentrated load of 1600 lb 3 ft from the wall. Find the maximum deflection.

12. An S 10 × 35 standard beam is simply supported with a span of 20 ft and a load of 10 000 lb placed at the beam's midpoint. Calculate (a) the maximum deflection and (b) the deflection if the load were evenly distributed over the length of the beam.

13. A simply supported 4 × 8 timber beam has a span of 12 ft. Two 700 lb concentrated loads are placed 3 ft from each end. Determine the maximum deflection (a) when the 8 in. dimension is vertical and (b) when the 4 in. dimension is vertical.

14. A W 8 × 58 steel beam 30 ft long is used as an overhead traveling crane. The crane supports a load of 4 tons. Determine the maximum deflection when the hoist is at the midpoint of the beam.

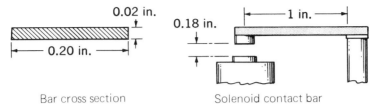

Bar cross section   Solenoid contact bar

FIGURE 17.6   Problem 9

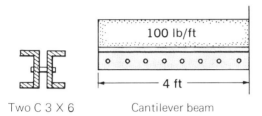

Two C 3 × 6   Cantilever beam

FIGURE 17.7   Problem 21

15. What center load can a 2 × 6 timber joist, with a 15 ft span, carry so that its maximum deflection is within the span/360 limit? This ceiling joist is simply supported.
16. A simply supported S 18 × 70 steel beam has a span of 40 ft. Find the maximum deflection due to its own weight if the web is placed (a) vertically and (b) horizontally.
17. (SI Units)   Find the maximum deflection of the beam in metric Sample Problem 17.6 if the load is evenly distributed over the length of the beam.
18. (SI Units)   Find the maximum deflection of a W 8 × 21 simply supported steel beam 10 m long with a center load of 3 Mg.
19. (Extra Credit)   A simply supported 4 × 12 timber beam has a span of 18 ft. A 4500 lb concentrated load is placed 5 ft from one end. Locate the position, and calculate the amount of maximum deflection.
20. (Extra Credit)   A solid steel rotating shaft 2 ft long takes a concentrated center load of 500 lb. The self-aligning bearings are placed at the ends of the shaft. The deflection limit for the shaft is 0.10 in. What shaft diameter is required?
21. (Extra Credit)   Two C 3 × 6 steel channels are bolted together to form a 4 ft cantilever beam as shown in Figure 17.7. Solve for the maximum deflection.
22. (Extra Credit)   Solve Sample Problem 17.2 for the maximum deflection, and include the weight of the beam.
23. (Extra Credit)   Solve Problem 8, and include the weight of the beam.

24. (Extra Credit) Solve Problem 11, and include the weight of the beam.

25. (Extra Credit) Solve Problem 13, and include the weight of the beam.

26. (Extra Credit) The American National Standard Building Code specifies that the minimum live load for exterior balconies is 100 lb·ft². (The live load is the load produced by the intended use of the balcony and does not include the weight of flooring or other structural material.) A balcony is to be designed to extend 8 ft from the building and to be 15 ft long. It will be supported by two evenly spaced cantilever standard steel S-beams. Deflection shall not exceed 0.2 in. Ignore the weight of the flooring and beams, and choose the lightest beam.

CASE STUDY

On a number of occasions during construction, beams for overpasses have been seated in place; but before being secured in position, they have toppled to the ground. In one such case, a high wind may have been involved. We must determine which of two conditions might apply:

1. The beam may have been cocked at an angle on its supports and fell down when it blew over.
2. When a W- or S-beam is turned on its side, the deflection may be so great as to cause it to slip through its supports.

Assume that the radius of curvature at the point of maximum deflection is the same for the whole beam; that is, the full length of the elastic curve is the arc of a circle. Determine the chord length of the elastic curve. Use this data: The beam is a W 10 × 112; its length is 50 ft; the support brackets are 5 in. deep.

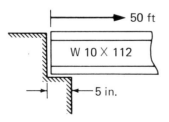

COMPUTER PROGRAM

The following program determines the maximum beam deflection for various combinations of beams and loads.

```
10   REM   DEFLECTION
20   PRINT "BEAM DEFLECTION PROGRAM. FOR A CANTILEVER WITH END LOAD-TYPE 1
     , FOR A CANTILEVER WITH"
21   PRINT "DISTRIBUTED LOAD-TYPE 2, FOR SIMPLE BEAM WITH CENTER LOAD-TYPE
     3, AND FOR"
22   PRINT "SIMPLE BEAM WITH DISTRIBUTED LOAD-TYPE 4."
30   INPUT N
40   PRINT
50   PRINT "TYPE IN ORDER: MODULUS OF ELASTICITY 'E',MOMENT OF INERTIA 'I'
     ,TOTAL LOAD (LB),"
51   PRINT "AND BEAM LENGTH (IN.)."
60   INPUT E,I,W,L
70   LET K = W / (E * I)
80   IF N = 1 THEN   GOTO 150
90   IF N = 2 THEN   GOTO 170
100  IF N = 3 THEN   GOTO 190
110  IF N = 4 THEN   GOTO 210
120  PRINT
130  PRINT "DEFLECTION = ";D;" IN."
140  END
150  LET D = K * (L ^ 3) / 3
160  GOTO 120
170  LET D = K * (L ^ 3) / 8
180  GOTO 120
190  LET D = K * (L ^ 3) / 48
200  GOTO 120
210  LET D = 5 * K * (L ^ 3) / 384
220  GOTO 120
```

18

Statically Indeterminate Beams

OBJECTIVES

This chapter demonstrates the concepts used in solving for reactions of statically indeterminate beams. In the process, you will become sufficiently skilled to solve for reaction of beams with the specific support and loading conditions discussed below.

INTRODUCTION

The phrase *statically indeterminate* refers to situations in which the three statics equilibrium equations, $\Sigma F_x = 0$, $\Sigma F_y = 0$, and $\Sigma M = 0$, are not sufficient to solve for the reactions supporting a load. The phrase does not mean that the problem is insoluble, because there are deflection formulas to help out. Chapter 9 introduced statically indeterminate axial loads; we will now concern ourselves with beam loadings. The statically indeterminate beams shown previously in Figure 13.1 are reproduced in Figure 18.1. This chapter deals with the three situations illustrated:

1. A beam fixed at one end and supported at the other.
2. A two-span continuous beam.
3. A beam fixed at both ends.

A *fixed beam* is a beam fastened so as to prevent its rotation. A cantilever beam is an example of a beam fixed at one end. *Supports* are considered to be pins, rollers, or knife edges that allow a beam to rotate. Notice the elastic curve in Figure 18.1(*a*). The beam is fixed in

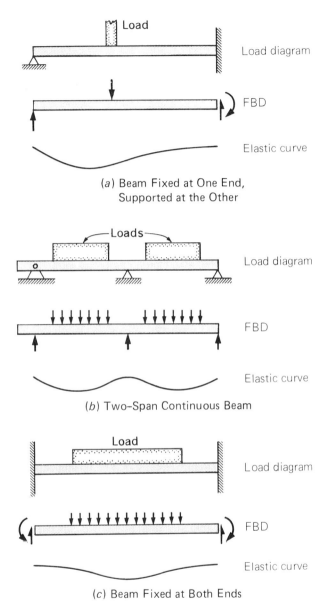

FIGURE 18.1 Selected Types of Statically Indeterminate Beams

the wall. That is, even though the beam deflects under load, it still must enter the wall at a horizontal angle. The wall does not allow movement of the beam. The knife edge, on the other hand, allows the beam to rotate away from the horizontal when loaded.

All solutions in this chapter are based on the assumption that the values for E and I are constant for the full length of the beam.

BEAM FIXED AT ONE END AND SUPPORTED AT THE OTHER

A beam fixed at one end and supported at the other is illustrated in Figure 18.1(a). In concept, when solving for reactions, the knife edge or roller support is removed, leaving a statically determinate cantilever beam. Since the end of the beam is now free to deflect, the deflection at the location of the support is determined. Holding this information aside, the next step is to construct a similar cantilever beam, the only load of which is applied at the support location. This force must be of such magnitude and direction as to deflect the beam an equal and opposite amount to the value of deflection determined in the first step. This force will be the support reaction in the original beam. The reactions at the wall can now be determined with the statics equations.

Actually, the deflection is not calculated. We simply set the formulas equal to each other. Figure 18.2 illustrates the procedure. It is interesting to note that the actual beam need not be specified since the modulus of elasticity and moment of inertia in each equation can be cancelled.

Procedure for Solving for Reactions at the Supported End of a Beam

Step 1 Write the equation for deflection of the beam at the support point with the support removed:

$$Y_1 = \frac{-A_{m1} \cdot \bar{x}_1}{E \cdot I}$$

Step 2 Write the equation for an equal and opposite deflection at the support location with all initial loads removed. Consider the support to offer the only applied force. With load P removed, apply R_A to obtain Y_2 equal and opposite to Y_1:

$$Y_2 = \frac{A_{m2} \cdot \bar{x}_2}{E \cdot I}$$

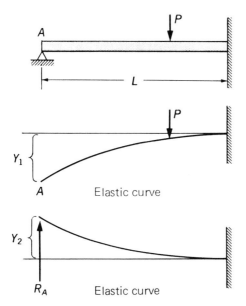

FIGURE 18.2 A Supported Cantilever Beam

Note that when solving for reactions, E and I drop out and, therefore, need *not* be calculated. Also, when solving for reactions, units of length drop out (leaving units of weight); therefore, lengths need *not* be in inches.

The beam in Figure 18.2 is given concrete loads and dimensions in Sample Problem 18.1.

SAMPLE PROBLEM 18.1
Loaded Beam Fixed at One End and Supported at the Other

Problem

Given the beam and load as sketched, determine reactions A and C.

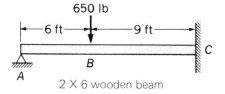

330 CHAPTER EIGHTEEN

Solution

Step 1 Support A is removed, and the deflection formula is set up.

$$Y_1 = \frac{-A_{m1} \cdot \bar{x}_1}{E \cdot I} = \frac{-1/2(650 \text{ lb} \cdot 9 \text{ ft} \cdot 9 \text{ ft})(6 \text{ ft} + 2/3 \cdot 9 \text{ ft})}{E \cdot I}$$

$$= \frac{-650 \cdot 9^2 \cdot 6 \text{ lb·ft}^3}{E \cdot I}$$

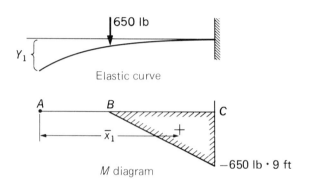

Step 2 Apply force R_A.

$$Y_2 = \frac{A_{m2} \cdot \bar{x}_2}{E \cdot I} = \frac{1/2(R_A \text{ lb} \cdot 15 \text{ ft} \cdot 15 \text{ ft})(2/3 \cdot 15 \text{ ft})}{E \cdot I}$$

$$= \frac{R_A \cdot (15)^2 \cdot 5 \text{ lb·ft}^3}{E \cdot I}$$

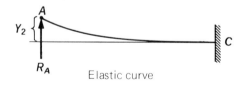

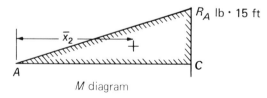

Step 3 $Y_2 = -Y_1$. The $E \cdot I$'s cancel out; therefore,

$$R_A \cdot (15)^2 \cdot 5 = -(-650 \cdot 9^2 \cdot 6)$$

$$R_A = \frac{650 \cdot 9^2 \cdot 6}{(15)^2 \cdot 5} = 281 \text{ lb} \uparrow$$

$\Sigma F_y = 0 = +281 - 650 + R_C$

$R_C = +369 \text{ lb} \uparrow$

$\Sigma M_C = 0 = +281 \cdot 15 - 650 \cdot 9 \pm M_C$

$M_C = +1635 \text{ lb·ft}$

MOMENT DIAGRAM BY PARTS

Sample Problem 18.2 has more complicated loadings and support locations. In such a situation, the calculations may be made easier if the moment diagram is broken down into its parts. That is, draw a separate moment diagram for each force or load, make the required calculations for each diagram, and then add the results.

SAMPLE PROBLEM 18.2
Beam Fixed at One End and Overhanging the Support

Problem

Given the beam and load as sketched, find reaction B.

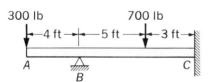

Solution

Step 1 Determine the deflection of the beam at B if support B is removed.

$$EIY_1 = A_m \cdot \bar{x} = -\left(300 \text{ lb} \cdot 4 \text{ ft} \cdot 8 \text{ ft} \cdot \frac{8}{2} \text{ ft}\right)$$

$$- \frac{1}{2}(300 \text{ lb} \cdot 8 \text{ ft} \cdot 8 \text{ ft})\left(\frac{2}{3} \cdot 8 \text{ ft}\right)$$

$$- \frac{1}{2}(700 \text{ lb} \cdot 3 \text{ ft} \cdot 3 \text{ ft})\left(\frac{2}{3} \cdot 3 \text{ ft} + 5 \text{ ft}\right)$$

$$= -38\,400 - 51\,200 - 22\,100 = -111\,700 \text{ lb·ft}^3$$

332   CHAPTER EIGHTEEN

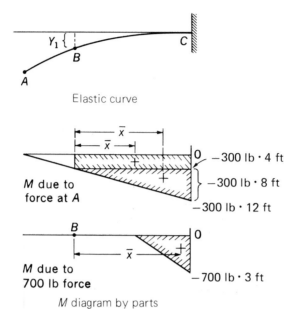

M diagram by parts

Step 2 Find the force at B necessary to deflect the beam equal and opposite to Y_1 (that is, $Y_2 = -Y_1$). No other loads are placed on the beam.

$$EIY_2 = A_m \cdot \bar{x} = \frac{1}{2}(R_B \cdot 8 \text{ ft} \cdot 8 \text{ ft})\left(\frac{2}{3} \cdot 8 \text{ ft}\right) = \frac{8^3}{3} \cdot R_B$$

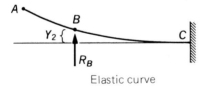

Elastic curve

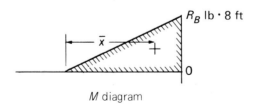

M diagram

Step 3 $EIY_2 = -EIY_1$. The $E \cdot I$'s cancel out; therefore,

$$\frac{8^3}{3} \cdot R_B = -(-111\,700)$$

$$R_B = 654 \text{ lb}$$

METRIC SAMPLE PROBLEM 18.3
Beam Fixed at One End and Supported at the Other

Problem

A cantilever 4 in. × 6 in. white pine beam 6 m long is supported at the free end and has a 600 kg load at its midpoint. Determine the free end support reaction in kilograms. Neglect the weight of the beam.

The required units in the formula $Y = (A_m \cdot \bar{x})/E \cdot I$ are Y, expressed in metres; A_m, in N·m²; $\bar{x}$, in metres. E and I will cancel out and need not be calculated.

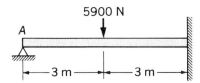

Solution

Step 1

$$Y_1 = \frac{-A_{m_1} \cdot \bar{x}_1}{E \cdot I}$$

$$-A_{m_1} = -\frac{1}{2}(17\,700 \cdot 3) = -26.6(10)^3 \text{ N·m}^3$$

$$\bar{x} = 3 + \frac{2}{3} \cdot 3 = 5 \text{ m}$$

$$Y_1 = \frac{-26.6(10)^3 \cdot 5}{E \cdot I} = \frac{-133(10)^3}{E \cdot I}$$

Elastic curve

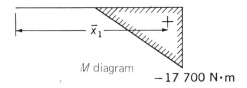

M diagram   −17 700 N·m

Step 2

$$Y_2 = \frac{A_{m_2} \cdot \bar{x}_2}{E \cdot I}$$

$$A_{m_2} = \frac{1}{2}(6 \cdot R_A \cdot 6) = 18R_A$$

$$\bar{x} = \frac{2}{3} \cdot 6 = 4 \text{ m}$$

$$Y_2 = \frac{18 R_A \cdot 4}{E \cdot I} = \frac{72 R_A}{E \cdot I}$$

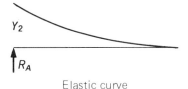

Elastic curve

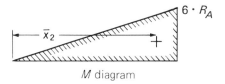

M diagram

Step 3 Set $Y_2 = -Y_1$, and the $E \cdot I$'s cancel out.

$$72 R_A = -[-133(10)^3]$$
$$R_A = 1.8(10)^3 \text{ N} = 190 \text{ kg}$$

TWO-SPAN CONTINUOUS BEAM

Reactions of the two-span continuous beam can be solved with an approach similar to that used for the cantilever beam fixed at one end and supported at the other end.

In concept, the middle support is removed to make the beam statically determinate and the deflection at the midpoint calculable. Next, a simple beam is constructed with a single concentrated load applied at the midpoint. This concentrated load must be sufficient to deflect the beam in an equal and opposite direction to that in the first step. This load is the center support reaction. Knowing the center reaction, the other reactions can be determined from the equations of statics. Since we are dealing only with loadings that are symmetrical and equal on each span, the end reactions are equal to each other, thus simplifying the solution.

In actual practice, the deflection, the moment of inertia, and the modulus of elasticity are not needed to find the beam reactions, as Figure 18.3 and the procedure given next indicate.

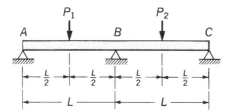

FIGURE 18.3 Two-Span Beam

Procedure for Solving for Reactions of a Two-Span Continuous Beam

Step 1 Remove the middle support, and write the equation for the deflection of the beam at this location. For the beam in Figure 18.3, remove support B. Because of symmetry and equal loads, we know that the maximum deflection will be at point B, and the tangent at B will be horizontal. The shaded area in the moment diagram below is used in the formula:

$$Y_1 = \frac{\Sigma A_m \cdot \bar{x}}{E \cdot I}$$

FBD and elastic curve

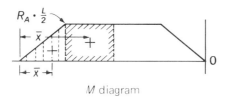

M diagram

Step 2 Remove the middle support and the given loads, and apply a single force at the midpoint of the beam to cause an equal and opposite deflection to that in Step 1. Write an equation for this deflection ($Y_2 = -Y_1$, or we can write $EIY_2 = -EIY_1$). The shaded area of the moment diagram is used in the formula:

$$Y_2 = \frac{A_m \cdot \bar{x}}{E \cdot I}$$

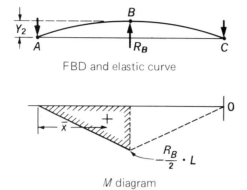

FBD and elastic curve

M diagram

Step 3 Set the two equations equal to each other to solve for the middle support reaction:

$$EIY_2 = -EIY_1$$

The beam in Figure 18.3 is given concrete loads and dimensions, and the reactions are solved in Sample Problem 18.4. Sample Problem 18.5 demonstrates the solution for a two-span continuous beam with a distributed load.

SAMPLE PROBLEM 18.4
Two-Span Continuous Beam with Concentrated Loads

Problem

Two equal and symmetrically placed loads are applied to a two-span continuous beam. Determine the reaction at B.

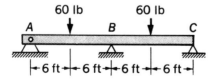

Solution

Step 1 Remove the center support. From past information, we know that location B has maximum deflection and a horizontal tangent. Also, by inspection, $R_A = R_C = 60$ lb.

Statically Indeterminate Beams   337

$$EIY_1 = \Sigma A_m \cdot \bar{x} = \frac{1}{2}(60 \text{ lb} \cdot 6 \text{ ft} \cdot 6 \text{ ft})\left(\frac{2}{3} \cdot 6 \text{ ft}\right)$$
$$+ (60 \text{ lb} \cdot 6 \text{ ft} \cdot 6 \text{ ft})(6 \text{ ft} + 3 \text{ ft})$$
$$= +23\,760 \text{ lb} \cdot \text{ft}^3$$

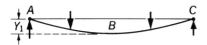

FBD and elastic curve

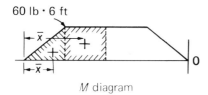

M diagram

Step 2 Apply a single load to beam at B so that deflection is equal and opposite to Y_1 ($Y_2 = -Y_1$, or $EIY_2 = -EIY_1$).

$$EIY_2 = A_m \cdot \bar{x} = -\frac{1}{2}\left(\frac{R_B}{2} \cdot 12 \cdot 12\right)\left(\frac{2}{3} \cdot 12\right) = -R_B \cdot 288$$

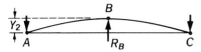

FBD and elastic curve

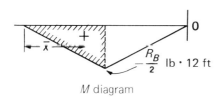

M diagram

Step 3 $EIY_2 = -EIY_1$. The $E \cdot I$'s cancel out; therefore,

$-R_B \cdot 288 = -(+23\,760)$
$R_B = 82.5 \text{ lb}$

SAMPLE PROBLEM 18.5
Two-Span Continuous Beam with a Distributed Load

Problem

Given the beam and distributed load as sketched, determine the reaction at B.

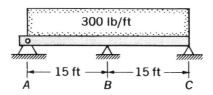

Solution

Step 1 Remove support B, and determine the deflection Y_1. By inspection, $R_A = R_C = 300 \cdot 15 = 4500$ lb. Therefore,

$$M_{max} = 4500 \cdot 15 \text{ ft} \cdot \frac{1}{2} = 3.37(10)^4 \text{ lb·ft}$$

The moment curve is a parabola; therefore, the shaded area is:

$$A_m = \frac{2}{3}[15 \text{ ft} \cdot 3.37(10)^4] = 3.37(10)^5 \text{ lb·ft}^2$$

$$\bar{x} = 0.6 \,(15 \text{ ft}) = 9.00 \text{ ft}$$

The area and $\bar{x}$ for a parabola were obtained from the Appendix.

$$EIY_1 = A_m \cdot \bar{x}$$
$$A_m \cdot \bar{x} = 3.37(10)^5 \cdot 9.00 = 3.03(10)^6 \text{ lb·ft}^3$$

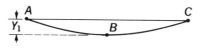

Elastic curve

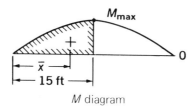

M diagram

Steps 2 and 3

Apply a force R_B so that $Y_2 = -Y_1$ $(EIY_2 = -EIY_1)$.

$$EIY_2 = A_m \cdot \bar{x} = -\frac{1}{2}\left(\frac{R_B}{2} \cdot 15 \cdot 15\right)\left(\frac{2}{3} \cdot 15\right) = -562R_B \text{ lb·ft}^3$$

$-562R_B = -3.03(10)^6$

$R_B = 5390$ lb

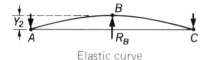

Elastic curve

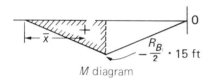

M diagram

BEAM FIXED AT BOTH ENDS

The solution of this type of beam requires the use of one of the moment-area formulas:

Formula 1: $\theta = \dfrac{A_m}{E \cdot I}$

Formula 2: $Y = \dfrac{A_m \cdot \bar{x}}{E \cdot I}$

Because the beam is fixed at both ends, as in Figure 18.1(c), the tangents to the elastic curve at either wall are horizontal and coincide. This means that the angle theta (θ) between the tangents at the walls is zero. In addition, if the loading is symmetrical, the tangent to the beam's midpoint is horizontal.

For beams with symmetrical loading, the quickest solution may be to apply moment-area Formula 1 between the wall and midpoint of the beam. Since the tangents are parallel, angle θ is again zero. Also, due to symmetrical loading, the shear reactions at the walls are equal. This leaves only the moment reactions at the walls to be determined.

Using the beam in Figure 18.4 as an example, the procedure

340 CHAPTER EIGHTEEN

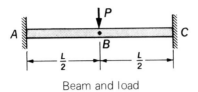

Beam and load

FIGURE 18.4 Beam Fixed at Both Ends

is simple: First construct the moment diagram and then write the appropriate formula to solve for the moment reaction at the walls. From the moment diagram, we see that because of symmetry, $V_A = V_C = P/2$, and $M_A = M_C$.

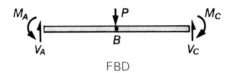

FBD

Since V_A and V_C can be determined easily, our problem is to solve for M_A. The elastic curve diagram shows that the tangents to the elastic curve at A and C are horizontal and coincide: $\theta_{AB} = 0$; $\theta_{AC} = 0$; and $Y_{AC} = 0$. The above can also be written as:

$$EI\theta_{AB} = A_m = 0$$
$$EI\theta_{AC} = A_m = 0$$
$$EIY_{AC} = A_m \cdot \bar{x} = 0$$

Note in the following formulas that M_A is the only unknown; V_A, P, and L are known. You may choose whichever formula suits you.

$$EI\theta_{AB} = A_m = \frac{1}{2}\left(\frac{V_A \cdot L^2}{4}\right) - \frac{M_A \cdot L}{2} = 0$$

$$EI\theta_{AC} = A_m = \frac{1}{2}(V_A \cdot L^2) - M_A \cdot L - \frac{1}{2}\left(\frac{PL^2}{4}\right) = 0$$

$$EIY_{AC} = A_m \cdot \bar{x} = \frac{V_A \cdot L^2}{2}\left(\frac{2}{3} \cdot L\right) - \frac{M_A \cdot L^2}{2}$$
$$- \frac{PL^2}{8}\left(\frac{L}{2} + \frac{2}{3} \cdot \frac{L}{2}\right) = 0$$

Statically Indeterminate Beams   341

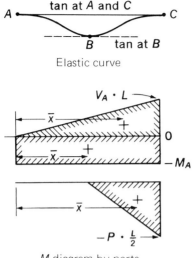

Elastic curve

M diagram by parts

Sample Problems 18.6 and 18.7 show the details in the solution of beams fixed at both ends and with symmetrical loads. In situations where the loading is not symmetrical, neither the shear forces nor the moments at the walls are equal to each other; therefore, both moment-area formulas must be set up and solved simultaneously. However, this text will discuss symmetrical loading only. The sample problems and problems at the end of the chapter all have symmetrical loads.

SAMPLE PROBLEM 18.6
Beam Fixed at Both Ends with a Center Load

Problem

Determine the moments at A and C.

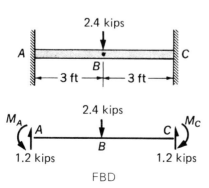

FBD

Solution

Because of symmetry, $V_A = V_C = 1.2$ kips and $M_A = M_C$. The tangent at A and C is parallel to the tangent at B; therefore, the angle θ between them is zero.

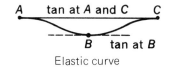

Elastic curve

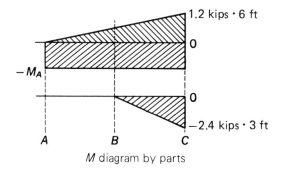

M diagram by parts

$EI\theta_{AB} = A_m = 0$

$A_m = \dfrac{1}{2}(1.2 \cdot 3^2) - M_A \cdot 3 = 0$

$M_A = \dfrac{0.6 \cdot 9}{3} = 1.8$ kip·ft

SAMPLE PROBLEM 18.7
Beam Fixed at Both Ends with Two Symmetrically Placed Loads

Problem

The beam fixed at both ends is loaded symmetrically as shown. Determine the moments at A and C.

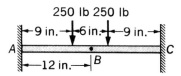

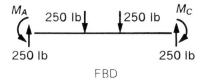
FBD

Solution

Because of symmetry, $V_A = V_C$ and $M_A = M_C$. By inspection, $V_A = 250$ lb. Note that the maximum beam deflection is at B, and the tangent at B is horizontal.

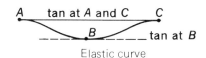
Elastic curve

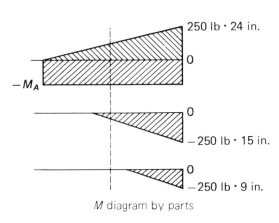
M diagram by parts

$EI\theta_{AB} = A_m = 0$

$A_m = \dfrac{1}{2}(250 \cdot 12^2) - M_A \cdot 12 - \dfrac{1}{2}(250 \cdot 3^2) = 0$

$M_A = \dfrac{125 \cdot 144 - 125 \cdot 9}{12} = 1406$ lb·in.

COMPLETE SHEAR AND MOMENT DIAGRAMS

You can draw complete shear and moment diagrams to aid in design of the beam, once all the beam reactions are determined, with beam weight neglected. Figure 18.5 portrays complete shear and moment

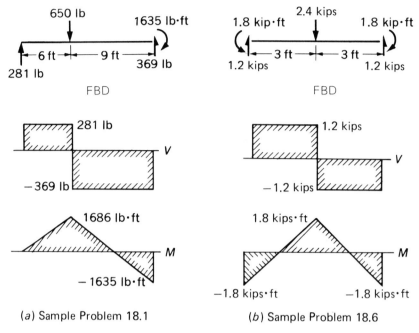

FIGURE 18.5 Complete Shear and Moment Diagrams

diagrams for the beams in Sample Problems 18.1 and 18.6. You know how to draw shear and moment diagrams, so instruction here is unnecessary. One word of caution, though! The area of the shear diagram up to any given position is not necessarily equal to the moment at that position. Look at Figure 18.5(b). There is an applied moment of 1.8 kip·ft at the left end that must be accounted for. The shear area may then be applied to this initial moment to determine other moment values.

SUMMARY

A beam fixed at one end and supported at the other and a two-span continuous beam are handled similarly. That is, one support is removed to make the beam statically determinate, and the formula for the deflection at the location of the removed support is written out. A second formula for an equal and opposite deflection caused by a single force applied at the support location is written, and these two formulas are set equal to each other. The single force is solved for and is equal to the missing support reaction. Once you obtain this information, you can solve the other support reactions in the usual manner with statics equations.

For a beam fixed at both ends, take advantage of the fact that the tangents to the elastic curve at the walls are horizontal. This means that the deflection of the beam, measured between the walls, is zero. Also, since the tangents are horizontal, the angle θ between the tangents will be zero. This gives us both moment-area deflection formulas to work with. If the load is placed symmetrically, the tangent to the elastic curve at the beam's midpoint will be horizontal. The formulas are:

Formula 1: $\theta = \dfrac{A_m}{E \cdot I}$

Formula 2: $Y = \dfrac{A_m \cdot \bar{x}}{E \cdot I}$

Once all reactions have been determined, the completed shear and moment diagrams for the statically indeterminate beam may be drawn. This will aid the process of choosing the proper size of beam.

PROBLEMS

Beams Fixed at One End and Supported at the Other

Determine all of the reactions in each problem. Consider the beams weightless unless otherwise specified.

1. A 30 in. cantilever beam has a support at one end and a 20 lb load at its midpoint.
2. A 12 ft beam has a 2000 lb load concentrated at its midpoint. A roller supports one end, and a wall provides support at the other end.
3. A cantilever beam with a support 9 ft from the wall supports a 3 kip load concentrated 6 ft from the wall.
4. A cantilever beam with a support 12 ft from the wall supports an evenly distributed load of 15 lb/ft.
5. A 40 in. cantilever beam has a 60 lb load concentrated at its free end. The beam overhangs its support by 10 in.; that is, the support is 30 in. from the wall.
6. (Extra Credit) A cantilever beam, length L, has a roller support distance L from the wall. A concentrated load P acts ver-

tically down on the midpoint of the beam. Write an equation for the roller reaction in terms of P and L.
7. (SI Units) A cantilever beam 3 m long, supported at its free end, holds a center load of 400 kg.

Two-Span Continuous Beams

Determine all of the reactions in each problem. Consider the beams weightless unless otherwise specified.

8. Each span is 6 ft long, and a 100 lb concentrated load is placed 3 ft from each of the outer ends of the spans.
9. Each span is 10 ft long, and a 2 kip concentrated load s placed at the midpoint of each span.
10. Each span is 8 ft long, and a load of 25 lb/ft is distributed evenly over both spans.
11. (Extra Credit) Each span has a length L and is center-loaded, with a concentrated load P. Write an equation for the middle support reaction in terms of P and L.
12. (SI Units) Each span is 9 m long, and a 500 kg concentrated load is placed at the midpoint of each span.

Beams Fixed at Both Ends

Determine the wall reactions. Consider the beams weightless.

13. Solve Sample Problem 18.6, making use of the facts that $EI\theta_{AC} = 0$ and $EIY_{AC} = 0$.
14. Solve Sample Problem 18.7, making use of the facts that $EI\theta_{AC} = 0$ and $EIY_{AC} = 0$.
15. A 60 in. long beam has a concentrated 700 lb load at its midpoint.
16. A 12 ft long beam has a concentrated 2000 lb load at its midpoint.
17. A beam with a 60 in. span has two concentrated loads. Each is 700 lb and placed 18 in. from the left and right walls.
18. A beam with a 12 ft span has three 500 lb concentrated loads placed 3 ft, 6 ft, and 9 ft from the right wall.
19. A 5 ft long beam has an evenly distributed load totaling 750 lb.
20. (Extra Credit) A beam with a span of L ft has a concentrated load P placed at its midpoint. Write an equation for the moment reactions in terms of P and L.
21. (SI Units) A beam with a 4 m span has an evenly distributed load totaling 1000 kg.

Complete Shear and Moment Diagrams

22. Draw the complete shear and moment diagrams for Sample Problem 18.2.
23. Draw the complete V and M diagrams for Sample Problem 18.4.
24. Draw the complete V and M diagrams for Sample Problem 18.5.
25. (SI Units) Draw the complete V and M diagrams for Problems 12 and 21.

COMPUTER PROGRAM

The following program will determine the three reactions supporting a two-span continuous beam. Spans must be equal in length, and the loads must be concentrated and placed symmetrically about the center support. Units are in feet and pounds.

```
10   REM  TWOSPAN
20   PRINT "TWO SPAN CONTINUOUS BEAM PROGRAM TO FIND SUPPORT REACTIONS.  L
     OADS (IN POUNDS)"
21   PRINT "MUST BE CONCENTRATED, EQUAL, AND PLACED SYMMETRICALLY ABOUT TH
     E CENTER SUPPORT."
30   PRINT
40   PRINT "TYPE IN ORDER: LOAD ON LEFT SPAN, THE DISTANCE (IN FEET) THIS
     LOAD IS FROM"
41   PRINT "THE LEFT END, AND THE LENGTH OF ONE SPAN IN FEET."
50   INPUT F,Z,L
60   LET Z1 = 0.3333 * F * (Z ^ 3)
61   LET Z2 = F * Z * (L - Z) * (Z + (L - Z) / 2)
62   PRINT
63   PRINT
70   LET B = (6 / (L ^ 3)) * (Z1 + Z2)
80   LET A = (2 * F - B) / 2
90   PRINT
100  PRINT "THE REACTION AT THE CENTER SUPPORT = "; INT (B * 10 + .51) /
     10;" LB."
110  PRINT
120  PRINT "THE REACTIONS AT EITHER END = "; INT (A * 10 + .51) / 10;" LB
     ."
130  END
```

19

Columns with Concentric Loads

OBJECTIVES

This chapter explains the theory of column failure and acquaints you with the various classes of columns, column terminology, and some of the handbook formulas. On completion of this chapter, you should be able to determine the allowable load when given the material, length, and shape of the column. In solving problems, you must determine the class of column, the appropriate formula, and whether or not a safety factor is required or already included in the formula. Also, you should be able to choose a satisfactory column to support a given load.

INTRODUCTION

Authorities believe the columns of our earliest civilizations had religious significance and were used to represent trees in the forest where the gods and spirits supposedly lived. As a technician, you will be interested in columns as load-bearing supports. You may have to consider loads on columns ranging in size from bridge supports to thumb tacks.

At one time or another, we have all probably placed a slender stick on end and pressed down to watch it bend (or buckle). If we were careful, the stick would return to its normal position when we removed the pressure. Let's take a similar situation and use an ordinary yardstick. Its cross-sectional dimensions are $1 \times 1/8$ in. A 1 in. length is cut from one end of the stick. This 1 in. portion is placed on end; we are asked to determine the load it can support. Use the formula:

$$\sigma = \frac{F}{A}$$

The handbook information in the Appendix gives 1400 psi for the allowable compressive stress. Since the cross-sectional area is 1/8 in.^{2}, the yardstick can support a compressive load of 175 lb as verified in a testing machine.

The next step is to take the 35 in. portion, stand it on end, and apply a compressive load with one finger. You know what will happen: The yardstick bends easily with a load much lighter than 175 lb. Another very interesting observation: If we are careful not to break the stick, it will straighten out when the load is removed. This means that even though the yardstick failed under load, the stresses never exceeded the elastic limit! Why does this happen, and how can safe loads for columns be determined? First, let's define some terms.

DEFINITIONS

Column: Columns are considered to be structural or machine members designed to take axial compressive loads.

Slenderness Ratio: The slenderness ratio is a term used to help simplify column formulas. The ratio is L/r. The L represents the unsupported length of the column, measured in inches, while r is the least radius of gyration of the column cross section. (The radius of gyration is explained in more detail in Chapter 5.) The radius of gyration is equal to $\sqrt{I/A}$, where I is the moment of inertia of the column cross section, and A is the cross-sectional area. For manufactured steel shapes, r is given in the data tables in the Appendix. The radius of gyration is not actually the outside radius of a column, but in order to picture the term *slenderness ratio* in our mind's eye, let us for a moment consider r to be the outside radius of a round column. In this situation, a column that is long and slim will logically have a larger slenderness ratio than one that is short and fat.

Column Failure: Column failure occurs when the column fails to support its load in the original position, regardless of whether or not the stresses have exceeded the elastic limit of the material.

Classes of Columns: Marks's *Standard Handbook for Mechanical Engineers* groups columns into three classes:

1. *Compression blocks* are columns short enough (slenderness ra-

tios generally below 30 for steel) so that the column will fail in compression before bending.
2. *Short columns* are members with slenderness ratios between the approximate values of 30 and 120 for steel. Other materials have somewhat different values.
3. *Long columns* are members with slenderness ratios greater than approximately 120 for steel. Bending is regarded as being the primary cause of failure in long columns.

Critical Load—Critical Stress: A long column is placed under a steadily increasing load in a testing machine. Some point will be reached where any attempt to increase the load is met by increased bending and the column may collapse. The load at this point is called the *critical load*, and the average stress on the cross section is called the *critical stress*. Most important, this critical load may be reached below the yield point of the material, which means that the column may fail while the stresses are still in the elastic range (as was demonstrated with the yardstick above).

Concentric Load: A concentric load means that the load is applied along the centroidal axis of the column. All loads in this chapter are assumed to be concentric.

Now that the more important definitions have been covered, let us examine the design of columns, starting with long columns.

LONG COLUMNS

The failure action of long columns follows a theoretical formula developed by the Swiss mathematician Leonhard Euler (1707–1783; pronounced *oi-lər*). Euler noted that the strength of a long column was not determined by the strength of the material but by the stiffness, and that a column is likely to deflect by bending when the load is not directly over the longitudinal axis. That is, the column behaves as if a load were deliberately placed eccentric to the longitudinal centroidal axis, which produces a couple that tends to bend the column, as illustrated in Figure 19.1. Note that as the column bends, the moment arm *e* increases; therefore, the bending moment may increase with no increase in load.

It appears a load can never be placed exactly over the longitudinal axis because:

Columns with Concentric Loads 351

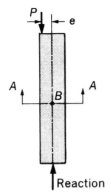

Assume the upper load P is placed eccentric to the longitudinal axis.

(a)

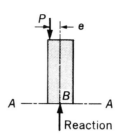

Since the load and reaction are not collinear, a counterclockwise couple is acting about point B:
$C = P \times e$.

(b)

As the column bending increases, the eccentricity e increases, causing the couple $P \times e$ to increase.

(c)

FIGURE 19.1 Euler's Concept of Column Behavior

1. The column cannot be made perfectly straight, but must be manufactured within limits. The American Institute of Steel Construction (AISC) specifications permit an initial bend of 1/1000 the length of a steel column, approximately 1/3 in. in a 25 ft column.
2. Inconsistencies in the material tend to set up localized stresses that may cause a moment to be applied about one of the cross section's centroidal axes.

3. Accidentally, the load may be placed eccentric to the longitudinal axis.

The derivation of Euler's formula is beyond the scope of this text, but we will discuss the concept. His formula is:

$$P_{cr} = \frac{n\pi^2 EA}{(L/r)^2}$$

where
P_{cr} = critical load
E = modulus of elasticity (the stiffness factor)
A = cross-sectional area of the column
L/R = slenderness ratio
n = coefficient to account for end conditions of the column

The values for n are given in the following paragraph.

Column End Conditions

Euler's formula is based on an analysis of round-ended columns. In this situation, the column is free to rotate or deflect about the contact points at the ends, but the ends are maintained in the same vertical line as shown in Figure 19.2. A column with one end free to rotate and one end fixed—that is, prevented from rotating as in Figure 19.2(b)—resists bending to a greater extent and appears to be stiffer. Therefore, a larger load can be supported than when both ends are free to rotate. When both ends are fixed, as shown in Figure 19.2(c), the column appears to be even stiffer. If one end is fixed and the other completely free to deflect, as is a flag pole or as shown in Figure 19.2(d), the column appears to be less stiff than one with rounded ends. To allow for these differences, the coefficient n mentioned above must be applied:

n = 1 when the column is rounded at both ends or pivoted at both ends

n = 2 when one end is fixed and the other end is rounded or pinned

n = 4 when both ends are fixed

n = 1/4 when one end is fixed and the other end is free

Average Stress

Because it is possible for a column to fail when stresses are well below the elastic limit, knowledge of the maximum bending stress in the

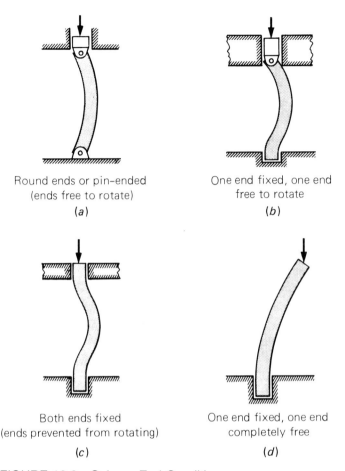

FIGURE 19.2 Column End Conditions

column is not as important as in beam design. Therefore, the *average stress* on the cross section may be used. The average stress on a column under critical load can be determined from Euler's formula simply by transposing the area A as follows:

$$\sigma_{cr} = \frac{P_{cr}}{A} = \frac{n\pi^2 E}{(L/r)^2}$$

The critical average stress on the column cross section is expressed as σ_{cr}.

Euler's formula is useful for those columns with large slenderness ratios. If the slenderness ratio falls to a value below about 120

TABLE 19.1 Typical Short Column Formulas

Formula	Material	Code	Slenderness Ratio
$\sigma_a = 17\,000 - 0.485(L/r)^2$	Carbon steels	AISC	$L/r < 120$
$\sigma_a = 16\,000 - 70(L/r)$	Carbon steels	Chicago	$L/r < 120$
$\sigma_a = 15\,000 - 50(L/r)$	Carbon steels	AREA	$L/r < 150$
$\sigma_{cr} = 34\,500 - (245/\sqrt{c})(L/r)$	2017 ST-aluminum	ANC	$[L/(\sqrt{c})r] < 94$
$\sigma_{cr} = 5000 - (0.5/c)(L/r)^2$	Spruce	ANC	$[L/(\sqrt{c})r] < 72$

σ_a = allowable (working) average stress
σ_{cr} = maximum (critical) average stress
c = 2 when both ends are pivoted (pin-ended)
 = 2.86 when one end is pivoted, one fixed
 = 4 when both ends are fixed
 = 1 when one end is fixed, one free

Note: Short column formulas may only be used with U.S. customary units.
AISC is the American Institute of Steel Construction.
AREA is the American Railroad Engineers Association.
ANC is the Army-Navy Civil Aeronautics Committee.

(for steel), other formulas must be used. These other formulas are discussed below; a few are listed in Table 19.1. The minimum slenderness ratio, below which Euler's formula may not be used, varies for different materials. Table 19.1 indicates these values for a few materials. Values for other materials may be obtained from appropriate handbooks such as Alcoa's *Structural Handbook* and Marks's *Standard Handbook for Mechanical Engineers*. Figure 19.3 compares the curve obtained from Euler's formula with actual failure data.

SHORT COLUMNS

Very few structures employ long columns; most structural columns fall in the short column classification. The average stress appearing on the cross section of a short column may be considered due partly to

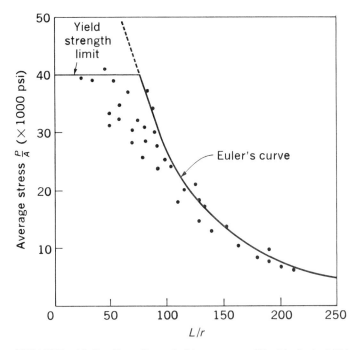

FIGURE 19.3 Results of Numerous Pin-Ended Mild Steel Columns

compression and partly to bending. A theoretical equation has not been derived. Empirical expressions are, in general, based on the assumption that the permissible stress must be reduced below that which could be permitted if the stress were due to compression only.

Typical formulas are represented in Table 19.1. The first three formulas cover structural steels and provide the average allowable or working stress. The term working stress is used by some handbooks to indicate allowable stress. There are two points to note about the first three formulas in the table:

1. As there is no lower cutoff value for the slenderness ratio, the formulas may also be used for compression blocks.
2. There is no coefficient to apply for end conditions, as in Euler's formula and in the last two formulas in the table. The first three formulas were designed for simplicity; since end conditions do not have as large an influence on short columns as on

long columns, the resulting slight variations in safety factor are acceptable.

Since 1961, the AISC has recommended a more sophisticated formula than the older AISC formula, which appears in Table 19.1 and in many handbooks. The newer formula does take into account column end conditions. More important, it takes into account the fact that a larger variety of steels are now used in construction. The formula is:

$$\sigma_a = \frac{[1 - (K \cdot L/r)^2/2C_c^2]\,\sigma_y}{5/3 + 3(K \cdot L/r)/8C_c - (K \cdot L/r)^3/8C_c^3}$$

The term σ_a is the allowable average stress on the column cross section.

A thorough study of the newer AISC formula with its restrictions and limitations is beyond the scope of this text. For those of you who wish to use the newer formula, the following conditions and values may be used:

$K = 1$ for columns pivoted at both ends

$C_c = 128$ for A36 steel

$\sigma_y = 36\,000$ psi (yield stress) for A36 steel

The last two formulas (ANC code) in Table 19.1 provide the critical average stress on the cross section of a column to which a safety factor must be applied. These two formulas also make use of a coefficient for end conditions as listed in the table.

Wood Column Formula

The ANC formula for spruce is not the most convenient formula to use in the construction industry. The Merritt's *Standard Handbook for Civil Engineers* provides the following formula for square or rectangular wood columns:

$$\sigma_a = \frac{0.30E}{(L/d)^2}$$

where

σ_a = allowable average compressive stress
E = modulus of elasticity
d = smallest cross-sectional dimension (actual, not nominal) when no bracing is involved

The formula applies to L/d ratios between 11 and 50. It was derived for pin-end conditions, but may also be used for square-cut ends. For round columns, the allowable stress may not exceed that for a square column of the same cross-sectional area.

Answers from the formula should not be used if they exceed the allowable stress for a compression block (L/d less than 11). The allowable compressive stress for compression blocks is listed in the Appendix. For L/d ratios greater than 50, use Euler's formula.

Safety Factor

The text *Elementary Mechanics of Deformable Bodies* by Smith and Sidebottom states that where a column formula provides the critical stress, a safety factor of 2 is recommended for structural grade steel, and a safety factor of 1.5 is recommended for aluminum alloy or stainless steel columns used in aircraft.

Figure 19.4 graphs the stress versus the slenderness ratio for a number of column formulas.

Procedure for Solving Column Problems

Step 1 To determine the load a given column can hold:
  (a) Calculate the L/r value, or the $L/(\sqrt{c})r$ value, if applicable.
  (b) From the information in Step (a), decide on the appropriate formula, and calculate the allowable stress. Use a safety factor if necessary.
  (c) With the allowable stress and the cross-sectional area of the column, determine the load.

Step 2 Choose a column to support a given load by trial, unless the column cross section is a simple geometric shape such as a square or circle, where the area changes in proportion to the radius of gyration.
  (a) We will assume that one of the short column formulas may be used and that the length of the column is known.
  (b) Choose the formula you will use. Set L/r equal to the maximum value allowed, and calculate the radius of gyration r.
  (c) Place the maximum L/r value in the formula you have chosen, and determine the cross-sectional area needed to support the given load—that is, the maximum area needed to support the load.
  (d) Determine the area needed if $L/r = 1$—that is, the minimum area needed to support the load.

358   CHAPTER NINETEEN

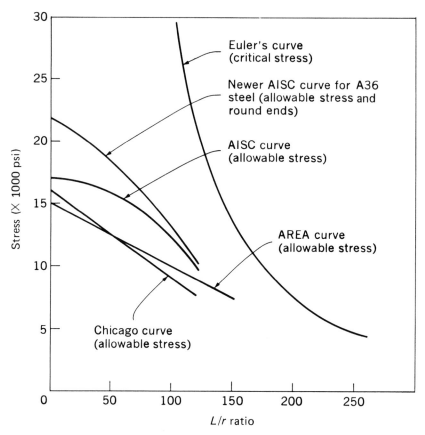

FIGURE 19.4   Curves of Stress vs. Slenderness Ratio for Several Short Column Formulas

(e) Go to the tables listing the properties of various beam and column shapes (in the Appendix and in various handbooks), pick a shape, and test it with the formula you have chosen. You know the cross-sectional area must be between the two values obtained in Steps (c) and (d). The radius of gyration r must be equal to or greater than the value obtained in Step (b). You can save yourself some work if you start with an area halfway between the limits you have established.

Sample Problems 19.1, 19.2, and 19.3 and metric Sample Problem 19.4 demonstrate the use of the column formulas.

SAMPLE PROBLEM 19.1
Slenderness Ratio

Problem

Find the slenderness ratio of (a) a 2 × 4 timber column 10 ft long, (b) an S 7 × 15.3 steel column 6 ft long, and (c) a solid round steel pushrod 1/4 in. in diameter × 5 in. long.

Solution (a)

Bending will take place about the Y-axis because the least radius of gyration is there.

$$r_y = \sqrt{\frac{I_y}{A}} = \sqrt{\frac{3.5(1.5)^3}{12 \times 3.5 \times 1.5}} = \sqrt{\frac{1.5^2}{12}} = 0.433 \text{ in.}$$

$$\frac{l}{r} = 10 \times \frac{12}{0.433} = 277$$

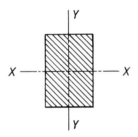

Solution (b)

From the table of properties of standard S shapes in the Appendix, the least $r = 0.766$ in.

$$\frac{L}{r} = \frac{6 \text{ ft} \times 12 \text{ in /ft}}{0.766} = 94.0$$

Solution (c)

From the Appendix, we note that the radius of gyration for a solid round cross section = diameter/4.

$$r = \frac{D}{4} = \frac{1/4}{4} = \frac{1}{16} = 0.0625 \text{ in.}$$

$$\frac{L}{r} = \frac{5}{0.0625} = 80$$

SAMPLE PROBLEM 19.2
Determining the Allowable Load

Problem

An aluminum box column is constructed as shown in the sketch. Both ends are fixed. Use Euler's formula or the ANC formula, whichever is applicable. Use a safety factor of 2. The column is 18 ft long. Determine the allowable load.

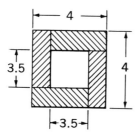

Solution

For a rectangular beam, $I = BH^3/12$. Since this cross section is square, B and H are equal. Therefore, $I = H^4/12$.

$$I_{\text{net area}} = \frac{H^4 - h^4}{12} = \left(\frac{1}{12}\right)(4^4 - 3.5^4) = 8.83 \text{ in.}^4$$

$$A_{\text{net}} = 16 - 12.22 = 3.78 \text{ in.}^2$$

$$r = \sqrt{\frac{I}{A}} = \sqrt{\frac{8.83}{3.78}} = 1.53 \text{ in.}$$

Check to see if $L/(\sqrt{c})r < 94$.

$$\frac{18 \text{ ft} \times 12 \text{ in./ft}}{\sqrt{4} \times 1.53} = 70.6$$

Use the ANC formula.

$$S_{cr} = 34\,500 - \frac{245}{\sqrt{4}}\left(\frac{L}{r}\right) = 34\,000 - 122.5\left(\frac{18 \times 12}{1.53}\right)$$

$$= 34\,500 - 17\,300 = 17\,200 \text{ psi}$$

$$\sigma_{\text{allow}} = \frac{17\,200}{2} = 8600 \text{ psi}$$

$$P_a = 8600 \times 3.78 = 32\,500 \text{ lb}$$

Columns with Concentric Loads 361

SAMPLE PROBLEM 19.3
Choosing a Column

Problem

A 15 ft column pinned at both ends supports a 170 000 lb load. Use the AISC code in Table 19.1, and choose the lightest weight W shape from the tables in the Appendix.

Solution

The maximum L/r is 120; therefore, the miminum r is (15 × 12 in./ft)/120, or 1.5.

$\sigma_a = 17\ 000 - 0.485(120)^2 = 10\ 000$ psi

maximum area required $= \dfrac{F}{\sigma_a} = \dfrac{170\ 000}{10\ 000} = 17$ in.^{2}

When L/r is 1,

$\sigma_a = 17\ 000 - 0.485(1)^2 = 17\ 000$ psi

minimum area required $= \dfrac{F}{\sigma_a} = \dfrac{170\ 000}{17\ 000} = 10$ in.^{2}

Choose a column with an area ≈ 13.5 in.^{2} and $r \geq$ 1.5 in. Choose W 10 × 45. $A = 13.24$ in.^{2} ; $r = 2.00$ in.

$P_a = \sigma_a \cdot A = \left[17\ 000 - 0.485 \left(\dfrac{15 \times 12}{2} \right)^2 \right] \cdot 13.24 = 173\ 000$ lb

This is very close and probably the answer. We cannot make any significant reduction in area unless we have a significant increase in r. We will check a lighter shape: W 8 × 40. $A = 11.76$ in.^{2}; $r = 2.04$ in.

$P_a = \sigma_a \cdot A = \left[17\ 000 - 0.485 \left(\dfrac{15 \times 12}{2.04} \right)^2 \right] \cdot 11.76 = 156\ 000$ lb

The lighter shape does not work; our answer is W 10 × 45.

METRIC SAMPLE PROBLEM 19.4
Long Column—Euler's Formula

Problem

Use Euler's formula

$\sigma_{cr} = \dfrac{n\pi^2 E}{(L/r)^2}$

and calculate the critical average stress on a round-ended steel column 10 m long, with a circular cross section diameter of 190 mm. The required units in the formula are σ_{cr}, in pascals; E, in pascals; L, in metres; r, in metres.

Solution

$$E = 207 \text{ GPa}$$

$$r = \frac{D}{4} \text{ (from Appendix)}$$

$$n = 1$$

$$\sigma_{cr} = \frac{1 \cdot \pi^2 (207)(10)^9}{[10/(0.19/4)]^2} = 464(10)^5 \text{ Pa} = 46 \text{ MPa (rounded)}$$

COMPRESSION BLOCKS

If no code applies, you may elect to use one of the short column formulas, or you may decide to use the basic strength of materials formula:

$$\sigma = \frac{F}{A}$$

You should use a safety factor suitable for the material and application.

SUMMARY

The slenderness ratio is very important in column formulas. You can visualize that the longer a column is compared to its radius of gyration (the larger the L/r ratio), the greater the influence of bending moments in causing column deflection and possible failure.

Euler's formula indicates that the strength of a material is unimportant in long columns, but stiffness and end conditions do affect the load a column can carry.

As the slenderness ratio decreases into the short column range, the strength of a material increases in importance, and the end conditions become less important.

Columns with Concentric Loads   363

PHOTO 19.1   Steel Columns and Diagonals of the Alcoa Building in San Francisco

The various formulas and their limitations can be confusing at first. The coefficient for end conditions in Euler's formula is represented by n, and the values listed for the ANC formulas are designated as c. The three steel formulas in Table 19.1 provide the allowable average stress. Euler's formula and the two ANC formulas provide the critical average stress to which a safety factor must be applied.

PROBLEMS

Unless otherwise specified, assume pin-ended (pivoted) conditions. When applying formulas that provide the critical stress, use a safety factor of 2. Use the AISC and ANC code formulas in Table 19.1 where applicable, unless otherwise specified.

1. Find the slenderness ratio of the following columns:
   (a) A solid round steel column 2 in. OD and 5 ft long.
   (b) A hollow round steel column 2 in. OD, with a wall thickness of 0.3 in., and a length of 5 ft.
   (c) A W 10 × 26 steel column 30 ft long.
   (d) A C 9 × 20 steel column 6 ft long.
2. Calculate the allowable loads for the columns in Problem 1. Use the AISC code in Table 19.1 where applicable.
3. Refer to Sample Problem 19.1. Determine the allowable loads for the three columns.
4. Find the allowable load for a W 8 × 21 steel column (a) 5 ft long, (b) 15 ft long, and (c) 30 ft long.
5. Determine the allowable load for a 4 × 8 timber column (a) 5 ft long, (b) 15 ft long, and (c) 30 ft long.
6. What is the allowable stress in an aluminum column 3 ft long and (a) 2 in. in diameter and (b) hollow with a 3 in. OD and 2 in. ID?
7. An S 10 × 35 steel shape 15 ft long is used as a column with one end fixed and one end free. What is the allowable load?
8. Four 2 × 4s are nailed together to form a square column, similar to the one in Sample Problem 19.2, intended to support a roof truss 12 ft above the floor. Consider one end fixed and the other end pivoted. Find the allowable load.
9. A flag pole 5 in. in diameter and 30 ft high is made of spruce. Could a 180 lb steeplejack climb this pole safely?
10. Push-type steel control rods 2 ft long and 1 in. in diameter are used on a large machine. What force can be applied to each rod? What force can be applied if the rods are made of aluminum? Both ends are fixed.
11. Columns supporting the upper head of a universal testing machine are 5 in. OD, steel, and 30 in. long. Consider one end fixed and the other free. Calculate the allowable load each column can support.
12. Choose the lightest W-shape column 25 ft long to support a load of 200 000 lb.
13. Choose the lightest S-shape steel column 8 ft long to support a 10 000 lb load.

14. Replace the steel column in Problem 13 with a spruce column with a square cross section. Determine the size.

15. (SI Units) Find the slenderness ratio for a round wood column 3 m long and 60 mm in diameter.

16. (SI Units) Use Euler's formula to calculate the critical stress on a solid brass column 2 m long and 40 mm in diameter.

17. (SI Units) An aluminum push rod in an engine is 10 mm in diameter and 150 mm long. Use the ANC code, and determine the critical load and the allowable load. Consider both ends of the push rod pinned.

COMPUTER PROGRAM

The AISC formula and Euler's formula are used in the following program to find the average stresses on a column. Data input is column length in feet and the least radius of gyration in inches.

```
10   REM   COLUMN FORMULAS
20   PRINT "AISC OR EULER'S STEEL COLUMN FORMULAS.  PROGRAM WILL PRODUCE A
     LLOWABLE AVERAGE STRESS FOR SHORT COLUMNS OR CRITICAL AVERAGE STRESS
     FOR LONG COLUMNS.  TYPE LENGTH IN FEET AND LEAST RADIUS OF GYRATION
     IN INCHES."
30   INPUT L,R
40   LET L1 = (L * 12) / R
50   PRINT
55   PRINT "THE SLENDERNESS RATIO ="; INT (L1 * 100 + .51) / 100
60   PRINT
70   IF L1 < 120 THEN  GOTO 300
80   FOR I = 1 TO 4
90   READ N(I)
100  NEXT I
110  DATA  1,2,4,.25
120  LET A$(1) = "BOTH ENDS ROUND."
130  LET A$(2) = "ONE END FIXED, ONE END ROUND."
140  LET A$(3) = "BOTH ENDS FIXED."
150  LET A$(4) = "ONE END FIXED, ONE END FREE."
155  LET E = 30000000
160  FOR J = 1 TO 4
170  LET S = N(J) * (3.1416 ^ 2) * E / (L1 ^ 2)
180  PRINT "THE LONG COL. CRIT. AVG. STRESS = "; INT (S * 10 + .51) / 10;
     " PSI FOR ";A$(J)
190  PRINT
200  NEXT J
210  END
300  LET S = 17000 - (0.485 * (L1 ^ 2))
310  PRINT "ALL. AVG. STRESS FOR THE SHORT COLUMN = "; INT (S * 10 + .51)
     / 10;" PSI."
320  PRINT
330  GOTO 210
```

Miscellaneous Topics

Mohr's Circle for Stress Determination (Forces Applied to Mutually Perpendicular Surfaces)

OBJECTIVES

This section explains how Mohr's circle is used to determine stresses on various planes in a machine member, when forces are applied to mutually perpendicular planes. On completion of this part of the chapter, you will be able to determine the angles of planes that have the maximum shear, tensile, and compressive stresses, and the stress values, given applied force or stress conditions on the machine member.

MOHR'S CIRCLE

As mentioned briefly in Chapter 8, two German engineers, Karl Culmann (1821–1881) and Otto Mohr (1835–1918), developed a method for solving stress problems that is now known as Mohr's circle. Culmann was an engineer and professor interested in graphic methods of solution. He developed a graphic method for stresses in a beam. Mohr, who was a highly regarded designer of railroad truss bridges, expanded and generalized Culmann's method to apply not only to structural members other than beams but also to other related problem areas such as strain determination.

In Chapter 8, Mohr's circle was used to find stresses caused by collinear force systems acting on structural members. The problems concerned stresses on planes at various angles to a member's cross section. You should review the information in Chapter 8, and then observe the bar in Figure 20.1. For our purpose of analysis, the forces

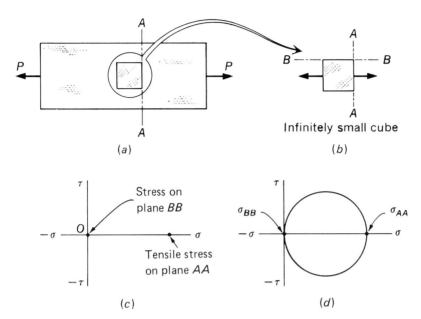

FIGURE 20.1 Relationship of Mohr's Circle to Stresses on Mutually Perpendicular Planes

must be reduced to stresses acting on an infinitely small cube in equilibrium. The cube must be infinitely small, because stresses that may be at right angles to each other must be considered as acting at a point. The concept of stresses acting at a point is important, because in some situations, such as in beams, the stresses change at different levels in the beam.

The stress on the right-hand plane of the cube is considered first; the plane is labeled AA in Figure 20.1(b). The stress on plane AA is plotted in Figure 20.1(c). There is no force, and therefore no stress, applied to the upper horizontal surface BB, which is at right angles to AA. This zero value is also plotted in Figure 20.1(c). Since the two surfaces of the cube are at right angles to each other, you will remember that their stresses are plotted diametrically opposite each other on Mohr's circle. Therefore, the distance between them forms the diameter of the circle depicted in Figure 20.1(d). From Mohr's circle, we can determine the tensile and shear stresses on planes at various angles to plane AA.

Let us now consider stresses due to forces applied on mutually perpendicular planes. With reference to Figure 20.2(a), note that a compressive force is applied to the horizontal planes. The problem is attacked as in the previous explanation. Determine the stresses on

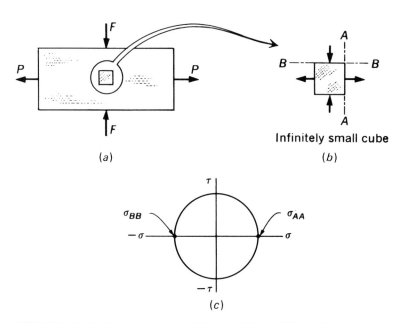

FIGURE 20.2 Construction of Mohr's Circle When Tensile and Compressive Forces Are Applied

planes AA and BB, plot these values, and construct Mohr's circle as in Figure 20.2(c). Note that compressive stresses are plotted to the left of the Y-axis. Compressive stresses are considered negative stresses, and the tensile stresses are considered positive stresses on the X-axis.

About all that is left now is to define the difference between positive and negative shear stresses in the Y direction. Figure 20.3 shows a member with shear forces applied on mutually perpendicular surfaces. We must visualize the front view of the small cube, as shown in Figure 20.3(b), as being able to pivot about an axis through its center. Engineers agree that if a shear stress on a given surface tends to rotate the cube clockwise, it will be labeled positive, and if rotation is counterclockwise, the shear stress is considered negative. Note that neither tensile nor compressive stresses are applied, but Mohr's circle indicates that tensile and compressive stresses are developed on planes other than AA and BB. Plane CC, 45° counterclockwise from AA (90° on Mohr's circle), has a compressive stress.

Procedure for Construction and Use of Mohr's Circle

Assume that forces acting on a machine or structural member are given or can be determined.

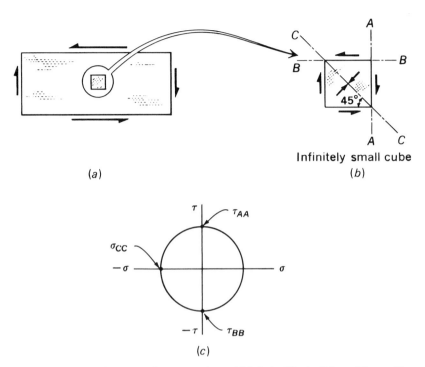

FIGURE 20.3 Construction of Mohr's Circle When Shear Forces Are Applied to Mutually Perpendicular Planes

Step 1 Sketch the infinitely small cube, align it to the given forces, and label its surfaces.

Step 2 Determine the applied tensile, compressive, or shear stresses on the mutually perpendicular surfaces of the cube.

Step 3 Set up the X- and Y-axes (written σ and τ) for the graph, and plot the stresses determined in Step 2. These stresses will determine coordinate points on the graph that are diametrically opposite each other on Mohr's circle.

Step 4 Calculate the location of the center of Mohr's circle. Since it falls on the σ axis, its location is the average of the tensile and/or compressive stresses. (This is demonstrated in Sample Problem 20.2.)

Step 5 Mohr's circle can now be constructed, and all of the required information can be taken from the circle through the use of trigonometric relationships.

Sample Problems 20.1 and 20.2 and metric Sample Problem 20.3 document the procedure for construction and use of Mohr's circle.

SAMPLE PROBLEM 20.1
Use of Mohr's Circle When Shear Forces Are Applied to a Machine Member

Problem

Given the information in the sketch, determine the plane and the corresponding stresses: (a) maximum shear stress, (b) maximum tensile stress, and (c) maximum compressive stress.

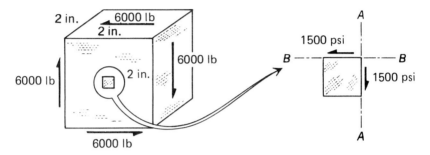

Step 1 First, locate the stresses applied to planes AA and BB on the stress axes.

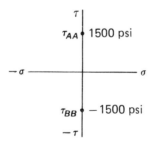

Step 2 Using the distance between the points as the diameter of Mohr's circle, construct the circle.

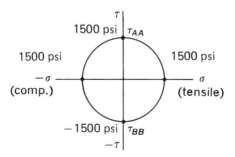

Solution (a)

By inspection, the maximum shear stresses are on planes AA and BB.

Solution (b)

The maximum tensile stress is 1500 psi on a plane 45° clockwise from plane AA.

Solution (c)

The maximum compressive stress is 1500 psi on a plane 45° counter-clockwise from plane AA.

SAMPLE PROBLEM 20.2
Use of Mohr's Circle for Stress Determination under Combined Loading

Problem

From the data given in the sketch, determine the planes and corresponding stresses: (a) maximum shear stress, (b) maximum tensile stress, and (c) maximum compressive stress.

Solution

Steps 1 and 2

The small cube is drawn, and the calculated stresses are added to the figure.

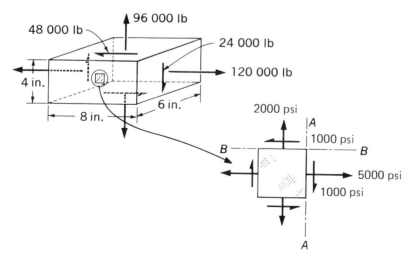

Step 3 The stresses are plotted.

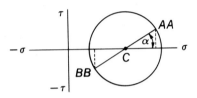

Step 4 The location for the center of Mohr's circle is determined.

$$\text{point } C = \frac{5000 + 2000}{2} = 3500 \text{ psi}$$

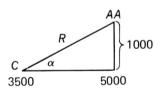

Step 5 Mohr's circle is drawn, and the required information is extracted from the data provided by the circle.

$$\text{angle } \alpha = \tan^{-1} \frac{1000}{1500} = 33.8° \qquad \frac{1}{2}\alpha = 16.9°$$

$$\text{angle } (90 - \alpha) = 56.2° \qquad \frac{1}{2}(90 - \alpha) = 28.1°$$

$$R = \sqrt{(1500)^2 + (1000)^2} = 1800 \text{ psi}$$

Solution (a)

The maximum shear stresses are equal to R, or 1800 psi, on planes 28.1° counterclockwise from planes AA and BB.

Solution (b)

Maximum tensile stress (3500 + 1800 = 5300 psi) is on a plane 16.9° clockwise from plane AA.

Solution (c)

No compressive stresses are developed.

METRIC SAMPLE PROBLEM 20.3
Use of Mohr's Circle for Tensile and Compressive Loads

Problem

An aluminum block 40 mm on a side has a vertical 3 kN tensile force system and a horizontal 4 kN compressive force system applied to it. Each force system consists of a force and its reaction. Determine the maximum tensile, compressive, and shear stresses.

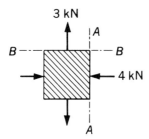

Solution

Mohr's circle is needed only for the determination of shear stress. Step 1 is omitted, since the applied stresses could be added to the original drawing, if needed.

Step 2

Surface *AA*: $\sigma_c = \dfrac{4(10)^3}{(0.04)^2} = -2.5$ MPa

Surface *BB*: $\sigma_t = \dfrac{3(10)^3}{(0.04)^2} = 1.9$ MPa

Step 4

point $C = \dfrac{+1.9 - 2.5}{2} = -0.3$ MPa

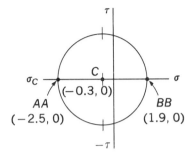

Step 5

circle radius = 1.9 + 0.3 = 2.2

radius = τ = 2.2 MPa

Eccentric Loads on Machine Members, Helical Springs, and Riveted Joints

OBJECTIVES

This part of the chapter explains the theory used in solving certain eccentric load problems, shows how the theory is modified to obtain formulas for practical applications, and demonstrates how the formulas are applied. On completion of this part, you will be able to choose the formula corresponding to the type of problem and will be able to apply the formula to solve for maximum stresses and loads.

COMPRESSION BLOCKS AND OFFSET AND CURVED MACHINE MEMBERS

In the previous chapters, we considered compression blocks and tension members loaded along a centroidal axis, as in Figure 20.4(a). Consideration will now be given to loads applied at some position other than along the centroidal axis, as in Figure 20.4(b). Note in the figure that

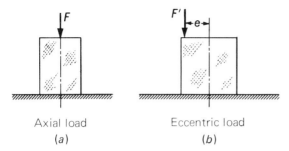

FIGURE 20.4 Loaded Compression Blocks Under Axial and Eccentric Loads

the distance from the centroidal axis to the load is called the *eccentricity* of the load, and the symbol is e.

Compression Blocks

The solution of an eccentric load problem is easily accomplished if the eccentric load is replaced with an equivalent load system consisting of a direct force applied along the centroidal axis and a couple. The method of obtaining this equivalent force system has been described in Chapter 3. You should review this chapter for a discussion of the methods and reasons whereby a single force can be replaced with a single force in another position and a couple.

The equivalent load system allows us to solve separately for the stress due to the direct force and the stress due to the couple. The results are superimposed in order to find the combined stress distribution. You have already solved direct force and couple problems in separate areas of this text; therefore, you need learn nothing new. Figure 20.5 diagrams the separate stresses and their combination.

The force F' in Figure 20.5(a) is replaced by (1) an equal force F along the centroidal axis and (2) a couple equal to the moment that the original force had about point A. This couple is expressed as $F \cdot e$. The new equivalent situation, shown in Figure 20.5(b), is tackled as two separate problems, because the single force F and the couple $F \cdot e$ can be solved separately for stress and the solutions superimposed. First, the stress due to the direct load F is determined by the formula: $\sigma = F/A$. This stress is shown graphically in Figure 20.5(c). The height H in the sketch (c) represents the magnitude of the stress. The second part of the problem deals with applied moments, which create bending

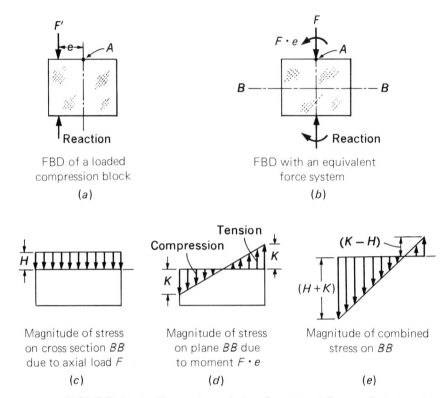

FIGURE 20.5 Illustration of the Combined-Stress Solution for Eccentric Loads

stresses just as in the beam problems studied in Chapter 14. Again, we are familiar with the bending stress formula:

$$\sigma = \frac{M \cdot c}{I}$$

This formula will be modified slightly for our solution; the only change being that M is replaced by $F \cdot e$, because this is the couple (moment) causing the stress. Of course, just as in beam problems, tensile and compressive stresses are developed. Figure 20.5(d) pictures the stress distribution due to the couple. The K in sketch (d) refers to the maximum bending stress values.

Now that the two sets of stresses have been determined, we superimpose one set of stresses on the other. Figure 20.5(e) shows the result. The compressive stresses resulting from the direct force and from the couple reinforce each other, and so are considered additive. The tensile stress of the couple opposes the compressive stress of the direct force; the difference between the stresses must therefore be determined. The formula that provides the maximum and minimum values of the stresses on a cross section is

$$\text{Eccentric load formula:} \quad \sigma = \pm \frac{F}{A} \pm \frac{(F \cdot e)c}{I}$$

The plus signs indicate tensile stress, and the minus signs indicate compressive stress.

Procedure for Solution of Eccentric Loads on Compression Blocks

Step 1 Resolve the eccentric load into a single force acting along the longitudinal axis of the member and a couple.

Step 2 Apply the formula above, using the correct sign, to determine the maximum tensile and compressive stresses.

Sample Problem 20.4 indicates the use of the above procedure.

SAMPLE PROBLEM 20.4
An Eccentrically Loaded Compression Block

Problem

From the information in the following sketch, determine the maximum compressive and tensile stresses, if any, on the cross section of the block.

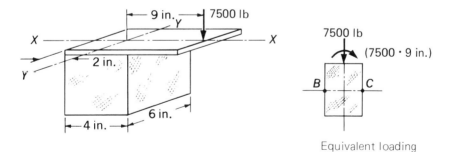

Equivalent loading

Solution

The moment is about the axis Y-Y; therefore,

$$I = \frac{6 \cdot 4^3}{12} = 32 \text{ in.}^4$$

378 CHAPTER TWENTY

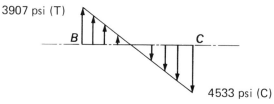

$$\sigma = -\frac{F}{A} \pm \frac{(F \cdot e) \cdot c}{I} = -\frac{7500}{4 \cdot 6} \pm \frac{(7500 \cdot 9)2}{32}$$

max $\sigma_{(tensile)} = -313 + 4220 = 3907$ psi (T)

max $\sigma_{(comp)} = -313 - 4220 = 4533$ psi (C)

Stresses on cross section B-C

An interesting situation presents itself when we consider an eccentrically loaded concrete or brick compression block. Masonry construction cannot resist a tensile stress for any practical application; therefore, an eccentric load on a masonry wall or block must be applied so that no tensile stress develops. Sample Problem 20.5 represents the approach to use.

SAMPLE PROBLEM 20.5
An Eccentrically Loaded Brick Post

Problem

A brick post in sketch (a) is to support an eccentric load of 30 000 lb. How far from the vertical centroidal axis can the load be placed without causing a tensile stress?

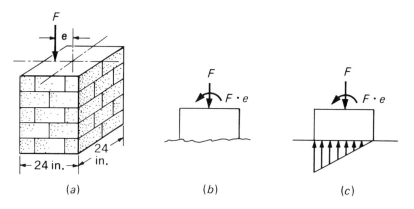

Solution

The equivalent loading is shown in (b), and the stress distribution must look like the distribution in (c).

Step 1 The compressive stress resulting from the direct force F:

$$\sigma_c = \frac{F}{A} = \frac{-30\,000 \text{ lb}}{576 \text{ in.}^2} = (-)52 \text{ psi}$$

↑ indicates compression

Step 2 The right-hand side of the brick post can have zero tensile stress; therefore, using the full formula:

$$\sigma_t = 0 = -52 \text{ psi} + \frac{(30\,000 \text{ lb} \times e \text{ in.}) \times 12 \text{ in.}}{27\,700 \text{ in.}^4}$$

$0 = -52 + 13 \times e$

$e = 4$ in.

The + indicates tensile stress on the right side of the post due to the couple.

Offset and Curved Machine Members

Offset machine members, such as those shown in Figures 20.17, 20.18, and 20.22, may be handled with the eccentric load formula above. However, not all curved machine members can be handled with this formula. E. P. Popov in his *Mechanics of Materials* text states that the eccentric load formula above, which includes the bending stress formula for straight beams, may be used for a curved beam whose radius of curvature is at least five times the depth of the cross section with no appreciable error. If the ratio is less than five, curved beam theory and formulas must be used, because the bending stress formula we use is based on the assumption that the neutral axis (axis of zero bending stress) coincides with the longitudinal centroidal axis; such is not the case in curved beams.

A complete discussion of curved beam theory is not within the scope of this text, but we can demonstrate why such a theory is needed. Figure 20.6(a) shows a straight rectangular beam deformed under load. Let the distance between two cross sections before loading be designated L. Since the original length L is the same at the top and bottom of the beam, a load causes the same strains on top and bottom. When the strains are equal, the stresses are equal, and the neutral axis coincides with the centroidal axis. Now let us consider the rectangular

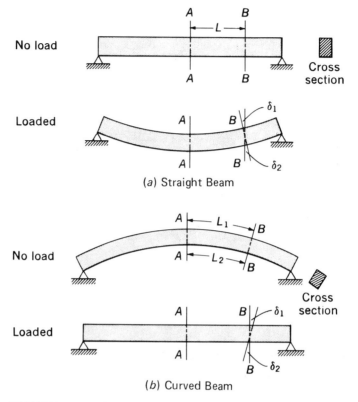

FIGURE 20.6 Comparison of Strains Developed in Straight and Curved Beams

curved beam in Figure 20.6(b). The loading is in such a direction as to straighten the beam. Before loading, the top and bottom distances between two cross sections are not the same length (L_1 is longer than L_2). This means that the strains will not be equal, even though the deformations (δ_1 and δ_2) are equal. If the top and bottom strains are unequal, the top and bottom stresses will be unequal, and the neutral axis will not coincide with the centroidal axis.

Procedure for Solution of Loads Applied to Offset and Curved Members

Step 1   Cut the member at a cross section on which the normal force is parallel to the applied force, and draw a free body diagram. This cross section should be the one that has the couple acting on it.

Miscellaneous Topics   381

Step 2 Follow the procedure outlined previously for applying the eccentric load formula.

Sample Problems 20.6 and 20.7 apply the solution procedure to a crane hook and a C-clamp.

SAMPLE PROBLEM 20.6
A Curved Crane Hook Problem

Problem

Data are given in the following sketch. If the allowable tensile or compressive stress is 18 000 psi, what is the allowable load F?

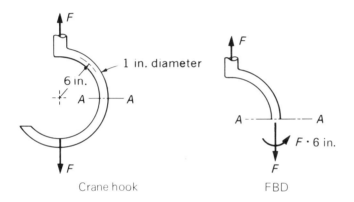

Crane hook   FBD

Solution

ratio of $\dfrac{\text{radial curve}}{\text{diameter of hook}} = \dfrac{6 \text{ in.}}{1 \text{ in.}} = 6$

The eccentric load formula can be used:

$$\sigma = \pm \frac{F}{A} \pm \frac{(F \cdot e)c}{I}$$

area of 1 in. diameter hook $= \pi \cdot \dfrac{1}{4} = 0.785 \text{ in.}^2$

I (for cross section of hook) $= \dfrac{\pi d^4}{64} = \dfrac{\pi \cdot 1}{64} = 0.049 \text{ in.}^4$

Step 1 We draw an FBD of the upper part of the hook. The FBD shows us that the maximum stresses will appear on the cross section at AA, because it is further away from the line-of-action of the applied load than any other cross section; and the inside edge of cross section AA has a combined tensile stress due to the direct force F and the couple F × 6. This combined stress at the inner edge will determine the applied load P, because the outer edge of AA has the tensile stress of the direct force opposing the compressive stress of the couple, resulting in a smaller combined stress.

Step 2 Apply the formula

$$+18\,000 = +\frac{F}{0.785} + \frac{F \cdot 6 \text{ in.} \cdot 1/2 \text{ in.}}{0.049}$$
$$= +127F + 61F = +62.27F$$
$$F = 288 \text{ lb}$$

METRIC SAMPLE PROBLEM 20.7
Eccentric Loads

Problem

The body of a C-clamp has a radius of 80 mm and a circular cross section diameter of 15 mm. A force of 3.0 kN is applied. What are the maximum tensile and compressive stresses in the clamp body?

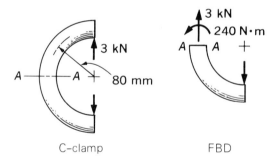

C-clamp   FBD

Solution

$$I = \frac{\pi D^4}{64} = \frac{\pi (0.015)^4}{64} = 2.5(10)^{-9} \text{ m}^4$$

$$A_{AA} = \frac{\pi D^2}{4} = \frac{\pi (0.015)^2}{4} = 177(10)^{-6} \text{ m}^2$$

ratio of $\frac{\text{radius}}{\text{diameter}} = \frac{80}{15} = 5.3$

The eccentric load formula can be used:

$$\sigma_t = \frac{F}{A} + \frac{(F \cdot e)c}{I} = \frac{3(10)^3}{177(10)^{-6}} + \frac{240(0.0075)}{2.5(10)^{-9}}$$

$$= 16.9(10)^6 + 720(10)^6 = 740 \text{ MPa (rounded)}$$

$$\sigma_c = 16.9(10)^6 - 720(10)^6 = 700 \text{ MPa (rounded)}$$

SHEAR STRESSES IN ROUND WIRE HELICAL SPRINGS

Investigation of shear stresses in the cross section of a helical spring requires a formula similar to the eccentric load formula. Figure 20.7(a) pictures a spring under load; Figure 20.7(b) is the free body diagram of a section of the spring. Note that the cross section of the coil must resist a force system consisting of a direct shearing force P and a couple, designated $P \cdot R$. The stresses developed by the force and the couple can be solved separately and then combined (superimposed). The direct shear stress on section AA in Figure 20.7(b) is equal to P/A. The shearing stress due to the applied couple is determined by an adaptation of the torsion formula:

$$\tau = \frac{T \cdot c}{J}$$

The torque symbol T is replaced by the symbol for the couple, which causes the torque. Therefore:

$$\tau = \frac{(P \cdot R)c}{J}$$

where

R = radius (in inches) of coil measured to centerline of wire
P = load, in pounds
c = radius of wire cross section, in inches
J = polar moment of inertia of wire cross section, measured in inches^4

The previous eccentric load formula can be adjusted for the shearing stress in helical springs. Thus:

Spring formula 1: $\quad \tau = \dfrac{P}{A} + \dfrac{P \cdot R \cdot c}{J}$

Only plus signs are used, because the maximum shearing stress de-

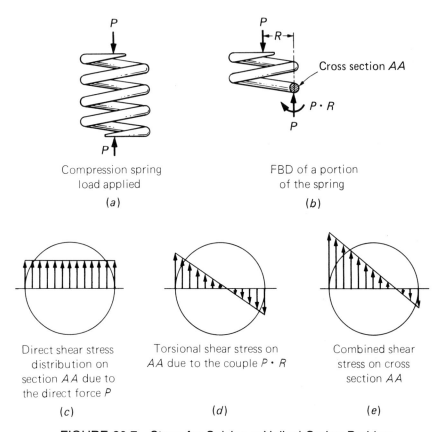

FIGURE 20.7 Steps for Solving a Helical Spring Problem

veloped is of interest, not the minimum stress. Parts (c), (d), and (e) of Figure 20.7 indicate the shear stress distribution on the cross section of a helical spring.

The spring formula above has a curvature limitation similar to that in the previous section on curved beams. The reason, depicted in Figure 20.8, is that the torsion formula was derived for a straight shaft. For the straight shaft in Figure 20.8(a), the deformation at A equals the deformation at B. The length is the same on both sides of the straight shaft; therefore, the strain at A equals the strain at B. This implies that the centroidal axis of the shaft has zero shear stress. However, note the curved shaft in Figure 20.8(b). Even though the deformation at A equals that at B, the strains will not be equal, because the lengths AC and BD are not equal. This means that the line of zero

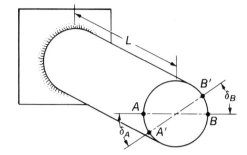

(a) Straight Shaft

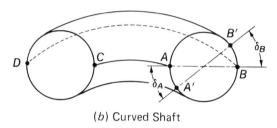

(b) Curved Shaft

FIGURE 20.8 Comparison of Shear Strains Developed in Straight and Curved Shafts

stress (neutral axis) does not coincide with the centroidal axis, and, therefore, the standard torsion formula does not work. For large ratios of R/c, the error is insignificant, but for smaller ratios it becomes important.

Mr. A. M. Wahl has performed experiments and developed a stress concentration factor K that greatly simplifies solutions to spring problems. Mr. Wahl's correction factor accounts for the error caused by the spring being a curved beam and for the direct shear stress, so that stresses in springs can be solved with his factor and the torsion formula:

$$\text{Spring formula 2:} \quad \tau = \frac{K \cdot P \cdot R \cdot c}{J}$$

K is the stress concentration factor and is graphed in Figure 20.9. It is plotted against the *spring index*, which is the term for the ratio R/c.

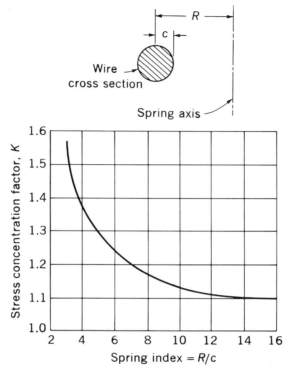

FIGURE 20.9 Stress Concentration Factor for Helical Round-Wire Compression or Tension Springs

Procedure for Solution of Loaded Spring Problems

Step 1 Calculate the spring index R/c.

Step 2 Go to Figure 20.9 with the spring index, and determine the concentration factor K.

Step 3 Substitute the known values in spring formula 2 and solve.

Sample Problem 20.8 indicates the usual approach for finding the allowable load.

SAMPLE PROBLEM 20.8
Springs

Problem

A helical spring has a wire diameter of 1/2 in. and a coil diameter (center to center of wire) of 6 in. It is made of spring steel with an allowable shear stress of 120 000 psi. What is the allowable load?

Solution

spring index $\dfrac{R}{c} = \dfrac{3}{1/4} = 12$

$K = 1.12$ (from Table 19.1)

$J = \dfrac{\pi r^4}{2} = \dfrac{\pi (1/4)^4}{2} = 0.00615$ in.^{4}

$\tau = \dfrac{K \times P \times R \times c}{J}$ (spring formula 2)

$P = \dfrac{\tau \times J}{K \times R \times c} = \dfrac{120\ 000 \times 0.00615}{1.12 \times 3 \times 1/4} = 878$ lb

ECCENTRICALLY LOADED RIVETED JOINTS

In a television story depicting his life, Christopher Columbus was somewhat put out at a party by guests belittling his accomplishments. Columbus asked if anyone could stand an egg on end. After everyone failed, he carefully crushed the shell so the egg would stand up. His point was that once the leaders show the way, others can follow. Of course, eccentrically loaded riveted joints present more of a problem than Columbus's demonstration, but actually there is very little to learn that is new. The leaders must show us how to assemble and place in sequence some of the concepts we have already mastered.

The concept used for solving the problems in the previous sections may be applied to the solution of eccentrically loaded riveted joints. That is, a single force may be replaced by an equivalent force system consisting of a direct force along a centroidal axis and a couple. In this way, the direct force on each rivet and the force due to the couple may be combined to obtain a single force on each rivet. Notice that we are now talking about combining forces, whereas in the previous section stresses were combined. There are two extra questions peculiar to riveted joints that must be answered prior to completing the solution:

1. Where is the centroid of a riveted joint?

2. In what ratio does the force vary from rivet to rivet due to the position of the individual rivet?

The answer to question 1 is that an eccentrically loaded riveted plate tends to rotate about a centroidal Z-axis of the rivet areas. This is important because the rivet areas (*not* the plate area) determine the centroid location. Figure 20.10 shows that the Z-axis is the centroidal axis perpendicular to the plate. Figure 20.10(*b*) indicates the free body diagram, with an equivalent force system consisting of the direct force F applied along the centroidal Y-axis of the rivet areas and a couple $F \cdot e$. As has been stated before, e is the eccentricity of the original load about the centroid.

There are two parts to the answer of question 2. There is a force on each rivet due to the direct force, and a force resulting from the action of the couple. The direct shear force on each rivet is easily computed, since the assumption is that each rivet takes its fair share of the total load. The force of the couple acting on each rivet is more difficult to determine. Rivets resisting a couple tend to deform, as indicated in Figure 20.10(*c*). It will be noticed that the shear deformation, and therefore the shear strain, is greater for rivets 1 and 4 than for rivets 2 and 3. Also, the magnitude of the strains is directly proportional to their distances from the centroid of the rivet areas (indicated as 0 in Figure 20.10). The reasoning is based on trigonometry, which states that the sides of any triangle have the same relationship to each other as the corresponding sides of a similar triangle. In Figure 20.10(*c*) the triangles 0-4-4' and 0-3-3' are similar, and side 4-4' is larger than side 3-3' in the same ratio as side 0-4 is to side 0-3, which means that the force on each rivet is proportional to its distance from the centroid of the rivet areas.

Remember that the torsion formula is useless here because the rivets are not being twisted (the same comment was made in Chapter 12 concerning the bolts in a bolted coupling). Instead, the action of the couple must be distributed to each rivet. The moment applied to each rivet is equal to a force on the rivet multiplied by the rivet's distance from the centroid. Stated algebraically:

$$F \cdot e = \Sigma f \cdot d$$

where
f = force on each rivet due to the applied couple
d = distance from each rivet to the centroid

The direction of the force on a rivet due to the couple is determined by the rotational direction of the couple. The line-of-action of the force is perpendicular to the line joining the rivet and the centroid.

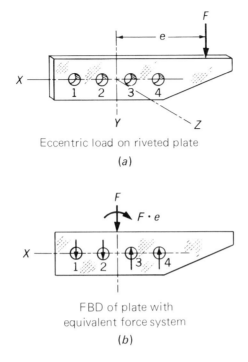
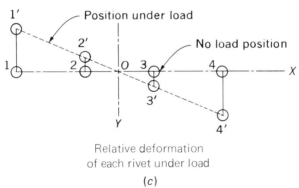

FIGURE 20.10 An Eccentrically Loaded Riveted Joint

Once the direct force and the couple force on each rivet have been determined, they are combined into a single force from which the stress is calculated by the basic formula $\tau = F/A$. A straightforward and concise formula for the combined forces on a rivet cannot be written as was done for machine members and springs. This is because the problem is complicated by varying distances and numbers of rivets and by the fact that if the rivets are not in a straight line, the total force on a rivet is the vectorial sum of the direct force and the couple force. Sample Problem 20.9 concerns a joint with rivets in a straight line; Sample Problem 20.10 deals with rivets in a triangular pattern.

Procedure for Solving Eccentrically Loaded Riveted Joint Problems

It is assumed all rivets attached to a given plate have the same cross-sectional area. Also, it is assumed loads are applied in the plane of the shear areas.

Step 1 Locate the position of the centroid of the rivet areas.

Step 2 Replace the original force with an equivalent force system consisting of a direct force acting through the centroid and a couple.

Step 3 Solve for the direct force on each rivet, and indicate by sketch the direction of each of these forces.

Step 4 Determine the magnitude and direction of the force on each rivet due to the couple. Indicate on a sketch. You should use the equation $F \cdot e = \Sigma f \cdot d$, and obtain the ratio of the force on each rivet compared to some selected single rivet to solve the equation. The sample problems following indicate how this is done.

Step 5 Combine the forces on each rivet vectorially to obtain a single force.

Step 6 Solve for the stress on each rivet if required.

SAMPLE PROBLEM 20.9
Solving Eccentrically Loaded Riveted Joints

Problem

Given the information in the sketch, solve for the combined shear stress on each rivet. Rivets are 1/2 in. in diameter. All dimensions are in inches.

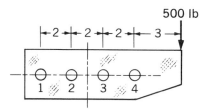

Solution

Step 1 The position of the centroid is determined by inspection.

Step 2 Set up an equivalent force system: 500 lb direct force down and a 3000 lb·in. clockwise couple.

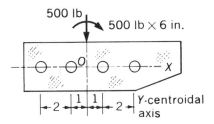

Step 3 Solve for the direct force on each rivet.

$$F = \frac{500}{4} = 125 \text{ lb} \downarrow$$

125 lb 125 lb 125 lb 125 lb
↓   ↓   ↓   ↓
○   ○   ○   ○
1   2   3   4

Step 4 Solve for the couple force on each rivet.

$$F \cdot e = \Sigma f \cdot d = 3 \cdot f_1 + 1(f_2) + 1(f_3) + 3(f_4)$$

Also, from similar triangles, $f_1 = f_4$, $f_2 = f_3$, and $f_4/f_3 = 3/1$. Since all forces are $\geq f_3$, we can substitute in the equation above.

$500 \times 6 = 3(3 \cdot f_3) + 1(f_3) + 1(f_3) + 3(3 \cdot f_3)$

$3000 = 20F_3$

$F_3 = F_2 = 150 \text{ lb}$

$F_1 = F_4 = 3(150) = 450 \text{ lb}$

The directions of the couple forces are shown in the sketch below. Note that they tend to produce clockwise rotation.

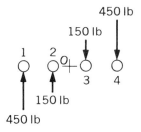

392 CHAPTER TWENTY

Step 5 Combine the forces.

Rivet 1: -125 lb $+ 450$ lb $= 325$ lb ↑
Rivet 2: -125 lb $+ 150$ lb $= 25$ lb ↑
Rivet 3: -125 lb $- 150$ lb $= 275$ lb ↓
Rivet 4: -125 lb $- 450$ lb $= 575$ lb ↓

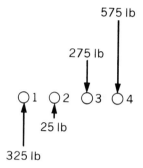

Step 6 Find the shear stress on each rivet with the formula $\tau = F/A$.

Rivet 1: $\tau = \dfrac{325}{0.196} = 1660$ psi

Rivet 2: $\tau = \dfrac{25}{0.196} = 128$ psi

Rivet 3: $\tau = \dfrac{275}{0.196} = 1406$ psi

Rivet 4: $\tau = \dfrac{575}{0.196} = 2940$ psi

SAMPLE PROBLEM 20.10
An Eccentrically Loaded Riveted Joint

Problem

Solve for the combined shear stress on each rivet in the sketch. The rivets are 1/4 in. in diameter; dimensions are in inches.

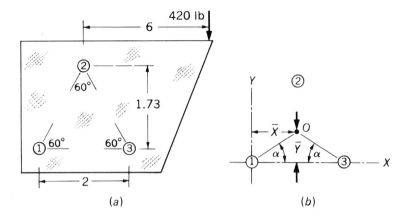

(a)   (b)

Solution

Step 1 Determine the position of the centroid from the reference axes in sketch (b).

$$\bar{X} = \frac{\Sigma ax}{\Sigma a} = \frac{\overset{\text{rivet 1}}{(0.05 \cdot 0)} + \overset{\text{rivet 2}}{(0.05 \cdot 1)} + \overset{\text{rivet 3}}{(0.05 \cdot 2)}}{0.15} = 1 \text{ in.}$$

$$\bar{Y} = \frac{\Sigma ay}{\Sigma a} = \frac{(0.05 \cdot 0) + (0.05 \cdot 1.73) + (0.05 \cdot 0)}{0.15} = 0.576 \text{ in.}$$

$$\alpha = \tan^{-1} \frac{0.576}{1} = 30°$$

Step 2 Set up an equivalent force system: 420 lb direct force down and a 2520 lb·in. clockwise couple.

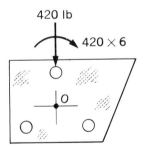

Step 3 Solve for the direct force on each rivet.

$$F = \frac{420}{3} = 140 \text{ lb} \downarrow$$

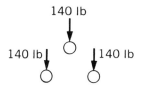

Step 4 Determine the magnitude, direction, and line-of-action of the force on each rivet due to the couple. The direction of the forces is clockwise. Lines-of-action are perpendicular to the line joining centroid and rivet. The distance from the centroid to each rivet is 1 in./cos 30°, or 1.155 in. Since the rivets are equidistant from the centroid, each force is the same.

$F \cdot e = \Sigma f \cdot d$

2520 lb·in = $3(f \cdot 1.155)$

f = 726 lb on each rivet

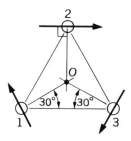

Step 5 Combine the forces (vectorial summation is required).

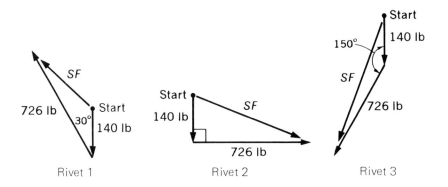

Cosine law: $c^2 = a^2 + b^2 - 2ab \cos C$

$(SF)^2 = (140)^2 + (726)^2 - 2(140)(726)(\cos 30°) = 371\ 600$

$SF_1 = \sqrt{371\ 600} = 609$ lb

$(SF)^2 = (140)^2 + (726)^2 = 547\ 600$

$SF_2 = \sqrt{547\ 600} = 738$ lb

Cosine law: $c^2 = a^2 + b^2 - 2ab \cos C$

$(SF)^2 = (140)^2 + (726)^2 - 2(140)(726)(\cos 150°)$
$= 546\ 600 - 203\ 000(-\cos 30°)$

$SF_3 = \sqrt{723\ 600} = 850$ lb

Step 6 Find the shear stress on each rivet with the formula $\tau = F/A$.

Rivet 1: $\tau = \dfrac{609}{0.05} = 12\ 180$ psi

Rivet 2: $\tau = \dfrac{738}{0.05} = 14\ 760$ psi

Rivet 3: $\tau = \dfrac{850}{0.05} = 17\ 000$ psi

SUMMARY

Mohr's circle is very helpful in solving problems dealing with forces applied to mutually perpendicular surfaces of a machine or structural member. It enables the technician and engineer to visualize the stresses developed on various planes in a member and reduces complicated formulas to simple trigonometric relationships.

When considering eccentric load problems, you will find the most practical approach to be that of separating the problem into two parts: one part considering stress (or force) due to direct centroidal loading and the other considering stresses due to a couple. The stresses are then superimposed on each other to obtain a combined stress value. This is demonstrated by the eccentric load formula for compression blocks and offset and curved machine members:

$$\sigma = \pm \dfrac{F}{A} \pm \dfrac{(F \cdot e)c}{I}$$

Frequently the combination of stresses is modified in some practical formulas to facilitate solution. This approach is represented by

Spring formula 2: $\tau = \dfrac{K \cdot P \cdot R \cdot c}{J}$

Solution of riveted joints with eccentric loads requires the concept of combining the direct force on each rivet with the force resulting from couple action.

PROBLEMS

Mohr's Circle

1. Given the information in Figure 20.11, determine the location of the planes and their corresponding maximum tensile, compressive, and shear stresses.
2. Repeat Problem 1, but use the information in Figure 20.12.

For Problems 3 through 7, sketch the small cube on which the stresses are acting, sketch Mohr's circle, and determine the location of the planes with the corresponding maximum tensile, compressive, and shear stresses. Refer to the small cube and planes sketched in Figure 20.1(b).

3. A tensile stress of 10 000 psi is applied to plane AA.
4. Shear stresses of 5000 psi are applied to planes AA and BB. The shear stress on AA is down in direction, and the shear stress on BB is directed to the left.
5. A 5000 psi tensile stress is applied to plane AA, and a 10 000 psi compressive stress is applied to plane BB.
6. Tensile stresses of 6000 psi act on AA and BB.
7. Plane AA has a 5000 psi tensile stress and 5000 psi shear stress acting downward. Plane BB has a 10 000 psi compressive stress and a 5000 psi shear stress acting to the left.
8. Refer to Figure 20.13. Draw Mohr's circle, and determine the planes and corresponding stresses for the conditions of maximum tensile stress, maximum compressive stress, and maximum shear stress.
9. Follow the instructions in Problem 8, but use the data in Figure 20.14.
10. (SI Units) A compression block 30 mm on a side has a 2 kN compressive force applied to the top horizontal surface. A 15 kN compressive force is applied to the vertical surface. Draw Mohr's circle, and determine the maximum compressive, tensile, and shear stresses, if any, and specify the surfaces on which they act.
11. (Extra Credit) A 2 in. diameter solid steel shaft resists a torque of 10 000 lb·in. Solve for the planes and the corresponding maximum stresses acting on a small cube on the surface of the shaft.

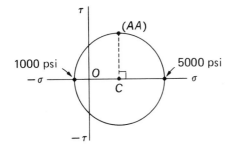

FIGURE 20.11 Problem 1

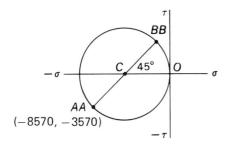

FIGURE 20.12 Problem 2

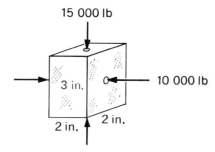

FIGURE 20.13 Problem 8

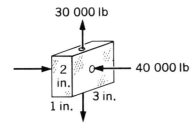

FIGURE 20.14 Problem 9

398 CHAPTER TWENTY

12. (Extra Credit) Solve for the planes and the corresponding maximum stresses acting on the small cube located as shown in the beam in Figure 20.15.

Compression Blocks and Offset and Curved Machine Members

For Problems 13 through 16, unless otherwise stated, solve for the maximum tensile and compressive stresses (if any). Sketch the stress distribution.

13. Refer to Figure 20.16. F is 108 000 lb; e is 1 in.
14. Refer to Figure 20.16. F is 108 000 lb; e is 4 in.
15. Refer to Figure 20.16. F is 108 000 lb; e is 6 in.
16. Refer to Figure 20.17. F is 108 000 lb; e is 12 in.
17. Given the information in Figure 20.18, determine the maximum stresses on section AA.
18. Refer to Sample Problem 20.5. If the load were (a) 20 000 lb and (b) 40 000 lb, would e change?
19. Figure 20.19 shows a square cement post with a load F of 18 000 lb. What is the maximum value for e, if no tensile stress is allowed to develop?
20. Figure 20.20 shows a curved C-clamp. If the applied force is 300 lb, determine the maximum tensile and compressive stresses on section AA.

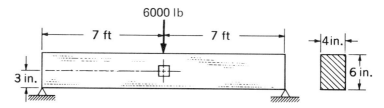

FIGURE 20.15   Problem 12

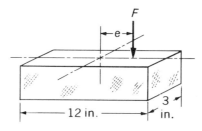

FIGURE 20.16   Problems 13, 14, and 15

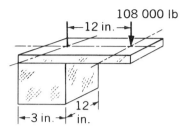

FIGURE 20.17 Problem 16

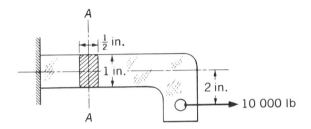

FIGURE 20.18 Problem 17

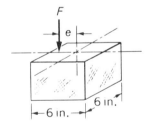

FIGURE 20.19 Problem 19

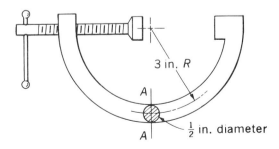

FIGURE 20.20 Problem 20

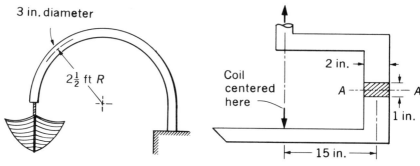

FIGURE 20.21 Problem 21   FIGURE 20.22 Problem 23

21. The lifeboat in Figure 20.21 exerts a force of 1000 lb on each davit. Determine the maximum tensile and compressive stresses developed in the davit, and identify the location of the corresponding cross section.
22. (SI Units) A circular concrete post 0.75 m in diameter has a concentrated compressive load of 10 kN placed 80 mm from the center of the post. Does this result in tensile stresses in the concrete?
23. (Extra Credit) Figure 20.22 illustrates a hook for carrying coils of sheet steel. If the allowable stress on the hook is 20 000 psi, tensile or compressive, what coil weight can be handled?
24. (Extra Credit) From the information you have gained in Sample Problem 20.5 and Problems 18 and 19, can you develop a rule or relationship between e and the dimensions of the cross section for the location of maximum eccentricity, if zero tensile stress is required?

Springs

25. (a) What stress is developed in a steel helical spring with a 1500 lb load, if the coil diameter is 5 in. and the wire diameter is 0.75 in.?
    (b) Solve (a), but substitute 3 in. for the coil diameter.
    (c) Solve (a), but substitute 0.50 in. for the wire diameter.
26. A phosphor bronze spring has an allowable shearing stress of 80 000 psi. What load can be applied if the coil diameter is 4 in. and the wire diameter is 1/2 in.?

27. What stress is developed in a steel helical spring with a load of 100 lb, a coil diameter of 1 in., and a wire diameter of 0.15 in.?
28. Determine the maximum coil diameter that will support a 5000 lb load if the wire diameter is 1½ in. and the allowable stress is 90 000 psi. This would make a good computer problem. If not using a computer, start at a spring index of eight, and move up or down in unit steps according to the value obtained.
29. (SI Units) What stress is developed in a steel spring with a 2.6 kN force, a coil diameter of 90 mm, and a wire diameter of 9 mm?

Eccentric Loads on Riveted Joints

30. Solve Sample Problem 20.9, but remove rivet 4. The first three rivets are to support the load.
31. Solve Sample Problem 20.9, but remove rivets 2 and 3.
32. Solve Sample Problem 20.9, but direct the 500 lb load downward midway between rivets 3 and 4.
33. Solve Sample Problem 20.9, but remove rivet 3.
34. Given the information in Figure 20.23, solve for the shear stress on each rivet.
35. Solve Sample Problem 20.10, but change the 2 in. distance between rivets to 4 in.

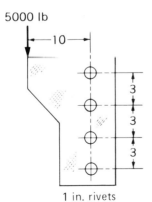

FIGURE 20.23   Problem 34

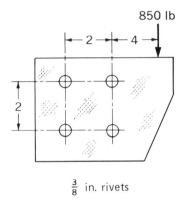

FIGURE 20.24 Problem 36

36. Given the information in Figure 20.24, solve for the shear stress on each rivet.

37. (SI Units) A bracket is held to a column by two rivets similar to the setup in Problem 31 above. Rivet centers are 200 mm apart, rivets have a cross-sectional diameter of 15 mm, and the applied force of 700 N is located between the rivets and 50 mm from the right rivet. Solve for the combined stress on each rivet.

COMPUTER PROGRAM

The maximum stresses appearing on a structural member can be determined with the following program. Their relative positions with respect to surface A are also calculated. Remember that tensile stresses are shown as + values and compressive stresses as − values. Shear stresses creating a CW twist are + and shear stresses creating a CCW twist are −. $S2$ must equal $S4$.

$S1$ = normal stress on surface A
$S2$ = shear stress on surface A
$S3$ = normal stress on surface B
$S4$ = shear stress on surface B
ST = maximum tensile stress
SC = maximum compressive stress
SS = maximum shear stress
A = angle theta on Mohr's circle

R = radius of Mohr's circle
C = distance of center of Mohr's circle from zero.

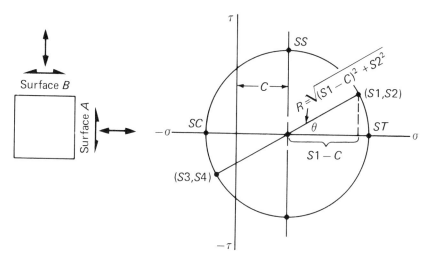

```
10   REM   MOHRTWO
20   PRINT "MOHR'S CIRCLE FOR COMBINED LOADING."
30   PRINT "TYPE IN ORDER: NORMAL STRESS ON SURFACE A, SHEAR STRESS ON A,
     NORMAL STRESS"
31   PRINT "ON B, SHEAR STRESS ON B."
40   INPUT S1,S2,S3,S4
50   PRINT
60   LET C = (S1 + S3) / 2
70   LET R = SQR (((S1 - C) ^ 2) + (S2 ^ 2))
75   IF S1 - C = 0 THEN GOTO 250
80   LET A = (180 / 3.1416) * ( ATN (S2 / (S1 - C)))
90   REM  "A IS THE ANGLE THETA ON MOHR'S CIRCLE"
100  LET ST = C + R
110  LET SC = C - R
120  LET SS = R
130  PRINT "THE MAXIMUM TENSILE STRESS = "; INT (ST * 10 + .51) / 10;" PS
     I."
140  PRINT
150  PRINT "THE MAXIMUM  COMP.  STRESS = "; INT (SC * 10 + .51) / 10;" PS
     I."
160  PRINT
170  PRINT "THE MAXIMUM  SHEAR  STRESS = "; INT (SS * 10 + .51) / 10;" PS
     I."
180  PRINT
190  PRINT " THE ANGLE THETA ON MOHR'S CIRCLE = "; INT (A * 100 + .51) /
     100;" DEG."
200  PRINT
210  PRINT "THE ANGLE BETWEEN SURFACE A AND THE ONE WITH THE MAX. TENSILE
      OR COMPRESSIVE STRESS = "; INT ((A / 2) * 100 + .51) / 100;" DEG."
220  END
250  LET A = 90
260  GOTO 100
```

Appendix: Handbook Data

METRIC (SI) SYSTEM GUIDE

The information below has been extracted from the ASME "Orientation and Guide for Use of Metric Units" and from information supplied by Dean C. Wandmacher, University of Cincinnati.

Angular measurements are in the conventional degrees (decimalized) and radians.

Note the spelling of *metre* and that people's names used as units are not capitalized except when in symbol form.

SI Unit Prefixes

Decimal Form	Exponent Form	Prefix Name	Prefix Symbol	Meaning
1 000 000 000	10^9	giga	G	billion
1 000 000	10^6	mega	M	million
1 000	10^3	kilo	k	thousand
0.001	10^{-3}	milli	m	thousandth
0.000 001	10^{-6}	micro	μ	millionth

SI Measurements

Conversion formulas: $A \cdot C = B$
$A = B/C$

(A) To convert	(B) to	(C) Multiply by
Length		
inches	metres (m)	0.0254
inches	millimetres (mm)	25.4
feet	metres (m)	0.305
Area		
square inches	square metres (m^2)	$6.45(10)^{-4}$
square feet	square metres (m^2)	0.0929
Volume		
cubic inches	cubic metres (m^3)	$16.4(10)^{-6}$
cubic feet	cubic metres (m^3)	0.0283
Weight		
pounds	kilograms (kg)	0.454
kilograms	newtons (N)	9.81
Force		
pounds	newtons (N)	4.45
Energy and Power		
foot·pounds	joules (J)*	1.36
horsepower	kilowatts (kW)**	0.746
Moment (Torque)		
pound·foot	newton·metre (N·m)	1.36
pound·inch	newton·metre (N·m)	0.113
Stress and Pressure		
pounds per square inch	pascals (Pa)†	6890.
pounds per square inch	newton/metre² (N/m²)	6890.
pounds per square inch	kilopascals (kPa)	6.89

*1 joule = 1 newton·metre
**1 watt = 1 joule/second
†1 pascal = 1 newton/metre²

COARSE-THREAD BOLT SIZES AND AREAS

This information is from Baumeister & Marks *Standard Handbook for Mechanical Engineers*, Eighth Edition. © 1978 by McGraw-Hill Book Company.

Diameter of Bolt in Inches	Number of Threads per Inch	Areas of Full Bolt in Square Inches	Areas of Bottom of Thread in Square Inches
1/4	20	0.049	0.027
3/8	16	0.110	0.068
1/2	13	0.196	0.126
5/8	11	0.307	0.202
3/4	10	0.442	0.302
1	8	0.785	0.551
1 1/4	7	1.227	0.890
1 1/2	6	1.767	1.294
1 3/4	5	2.405	1.745
2	4 1/2	3.142	2.300

THERMAL COEFFICIENT OF LINEAR EXPANSION (α)

This information is from LeGrande's *The New American Machinist's Handbook*. © 1955 by McGraw-Hill Book Company.

The thermal coefficient of linear expansion is the change in length, per unit of length, for a one degree change in temperature. Therefore, length units can be in inch/inch, foot/foot, metre/metre, etc.

Material	inch/inch/degree °F	°C
Aluminum alloy (average)	$0.000\,013 = 13(10)^{-6}$	$23(10)^{-6}$
Copper	$0.000\,0091 = 9.1(10)^{-6}$	$16(10)^{-6}$
Brass	$0.000\,011 = 11(10)^{-6}$	$20(10)^{-6}$
Bronze	$0.000\,010 = 10(10)^{-6}$	$18(10)^{-6}$
Gray cast iron	$0.000\,0058 = 5.8(10)^{-6}$	$10(10)^{-6}$
Steel (average)	$0.000\,0065 = 6.5(10)^{-6}$	$12(10)^{-6}$

ALLOWABLE STRESSES FOR STEEL AND ALUMINUM

The following allowable stresses for steel and aluminum have been extracted and simplified from data in the American Institute of Steel Construction's *Manual of Steel Construction,* Eighth Edition, and from Alcoa's *Structural Handbook.*

Stresses	A36 Steel		Aluminum Alloy Type 6061–T6	
	psi	MPa	psi	MPa
Axial tension and axial compression on short lengths	22 000	150	17 000	120
Bending tension and compression	24 000	160	17 000	120
Bearing on rivets and bolts	87 000*	600	30 000	210
Shear	14 400	99	11 000	76

*Experiments have shown that the strength of the connected part is not affected even when the bearing stress exceeds the ultimate tensile stress.

ALLOWABLE STRESSES FOR LUMBER

The information below has been selected and modified from a table of values in the *Standard Handbook for Mechanical Engineers,* by Baumeister & Marks. The values below may be considered typical for Douglas fir and southern pine, although there can be sizable variations in allowable stress between grades of lumber.

Compression parallel to the grain = 1400 psi (9.7 MPa)
Compression perpendicular to the grain = 400 psi (2.8 MPa)
Bending stress (compression or tension) = 1900 psi (13 MPa)
Shear stress parallel to the grain = 120 psi (830 kPa)

LUMBER SIZES

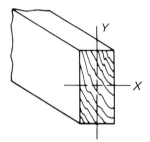

Location of axes on lumber cross section

Nominal Size	Actual Size (dry)	Cross-Sectional Area (in.^{2})	I_x* (in.^{4})	S_x** (in.^{3})
1 × 2	3/4 × 1 1/2	1.125	0.211	0.281
× 4	× 3 1/2	2.625	2.68	1.53
× 6	× 5 1/2	4.125	10.4	3.78
2 × 4	1 1/2 × 3 1/2	5.25	5.36	3.06
× 6	× 5 1/2	8.25	20.8	7.56
× 8	× 7 1/4	10.88	47.6	13.1
× 10	× 9 1/4	13.88	98.9	21.4
× 12	× 11 1/4	16.88	178	31.6
× 14	× 13 1/4	19.88	291	43.9
× 16	× 15 1/4	22.88	443	58.1
4 × 4	3 1/2 × 3 1/2	12.25	12.5	7.15
× 6	× 5 1/2	19.25	48.5	17.6
× 8	× 7 1/4	25.38	111	30.7
× 10	× 9 1/4	32.38	231	49.9
× 12	× 11 1/4	39.38	415	73.8
× 14	× 13 1/4	46.38	678	102
× 16	× 15 1/4	53.38	1030	136

Note: The average weight of lumber is 40 lb/ft^3.

*I_x is the moment of inertia of the cross-sectional area ($bH^3/12$) about the X-axis.

**S_x is the section modulus (I_x/c) of the cross section.

PROPERTIES OF MATERIALS

This information is from Baumeister & Marks *Standard Handbook for Mechanical Engineers*, Eighth Edition, © 1978 by McGraw-Hill Book Company and from ASME's *Handbook of Metal Properties*. The values are typical, unless otherwise noted.

Material	Ultimate Tensile Strength (Stress)		Yield Strength (Tensile or Compressive Stress)		Percent Elongation	Modulus of Elasticity E		Modulus of Rigidity G	
	psi	MPa	psi	MPa		psi	GPa	psi	GPa
Class 30 cast iron	30 000 (minimum) (115 000 comp) (44 000 shear)	207 793 303	—	—	—	15 000 000	103	5 000 000	34
ASTM A36 low-carbon structural steel	58 000 to 80 000	400 to 552	36 000	248	20	30 000 000	207	12 000 000	83
AISI 1020 cold drawn (0.2% carbon steel)	75 000	517 379	64 000	441	20	30 000 000	207	12 000 000	83
AISI 1020 hot worked	65 000	448	48 000	331	36	30 000 000	207	12 000 000	83
AISI 1040 cold drawn (0.4% carbon steel)	97 000	669	82 400	568	16	30 000 000	207	12 000 000	83
AISI 4340 oil quenched steel	212 000	1460	200 000	1378	12.5	30 000 000	207	12 000 000	83
AISI 302 stainless steel	90 000	621	40 000	276	50	28 000 000	193	10 000 000	69
Copper (soft)	32 000	221	10 000	69	45	15 600 000	108	6 000 000	41
Brass, 30% zinc (hard)	76 000	524	63 000*	434	—	16 000 000	110	6 000 000	41

(continued)

PROPERTIES OF MATERIALS (continued)

Material	Ultimate Tensile Strength (Stress)		Yield Strength (Tensile or Compressive Stress)		Percent Elongation	Modulus of Elasticity E		Modulus of Rigidity G	
	psi	MPa	psi	MPa		psi	GPa	psi	GPa
Brass, 30% zinc (soft)	47 000	324	15 000*	103	62				
6061–T6 aluminum alloy for structural shapes	45 000	310	40 000	276	17	10 600 000	73.0	4 000 000	28
Douglas fir (subjected to bending forces)	12 400	85.4	—	—	—	1 760 000	12.1	—	—

*0.5 percent extension under load

Average weight of steel = 0.284 lb/in.^{3}

Average weight of wood = 0.0231 lb/in.^{3}

PROPERTIES OF GEOMETRIC SHAPES

Shape	Area	c	I	S	J	r
Rectangle	BH	$\dfrac{H}{2}$	$\dfrac{BH^3}{12}$	$\dfrac{BH^2}{6}$		$\dfrac{H}{\sqrt{12}}$
Hollow rectangle	$BH - bh$	$\dfrac{H}{2}$	$\dfrac{BH^3 - bh^3}{12}$	$\dfrac{BH^3 - bh^3}{6H}$		
Circle	$\dfrac{\pi D^2}{4}$	$\dfrac{D}{2}$	$\dfrac{\pi D^4}{64}$	$\dfrac{\pi D^3}{32}$	$\dfrac{\pi D^4}{32}$	$\dfrac{D}{4}$
Hollow circle	$\dfrac{\pi}{4}(D^2 - d^2)$	$\dfrac{D}{2}$	$\dfrac{\pi}{64}(D^4 - d^4)$	$\dfrac{\pi}{32}\left(\dfrac{D^4 - d^4}{D}\right)$	$\dfrac{\pi}{32}(D^4 - d^4)$	$\dfrac{\sqrt{D^2 + d^2}}{4} $

(continued)

PROPERTIES OF GEOMETRIC SHAPES (continued)

Shape	Area	c	I	S	J	r
(half circle, radius R)	$\dfrac{\pi D^2}{8}$	$0.424R$				
(triangle, base B, height H)	$\dfrac{BH}{2}$	$\dfrac{H}{3}$	$\dfrac{BH^3}{36}$			

c = distance from centroid to indicated edge
I = moment of inertia about centroidal X axis
S = section modulus I/c
J = polar moment of inertia about centroidal Z axis
r = radius of gyration about centroidal X axis

PROPERTIES OF PARABOLAS

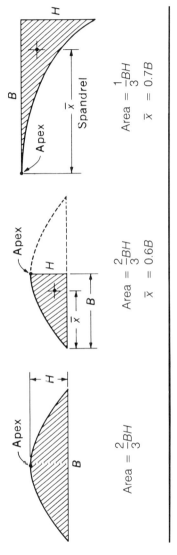

Information on parabolas is primarily for use with moment diagrams of beams.

PROPERTIES OF S SHAPES FOR DESIGN PURPOSES (STANDARD I BEAMS)

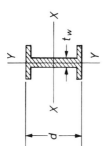

Designation	Area A (in.^{2})	Depth d (in.)	Web Thickness t_w (in.)	Axis X-X I (in.^{4})	Axis X-X S (in.^{3})	Axis X-X r (in.)	Axis Y-Y I (in.^{4})	Axis Y-Y S (in.^{3})	Axis Y-Y r (in.)	Nominal Weight per Foot (lb)
S 24 × 121	35.6	24.50	0.800	3160	258	9.43	83.3	20.7	1.53	121
× 106	31.2	24.50	0.620	2940	240	9.71	77.1	19.6	1.57	106
S 24 × 100	29.3	24.00	0.745	2390	199	9.02	47.7	13.2	1.27	100
× 90	26.5	24.00	0.625	2250	187	9.21	44.9	12.6	1.30	90
× 80	23.5	24.00	0.500	2100	175	9.47	42.2	12.1	1.34	80
S 20 × 96	28.2	20.30	0.800	1670	165	7.71	50.2	13.9	1.33	96
× 86	25.3	20.30	0.660	1580	155	7.89	46.8	13.3	1.36	86
S 20 × 75	22.0	20.00	0.635	1280	128	7.62	29.8	9.32	1.16	75
× 66	19.4	20.00	0.505	1190	119	7.83	27.7	8.85	1.19	66
S 18 × 70	20.6	18.00	0.711	926	103	6.71	24.1	7.72	1.08	70
× 54.7	16.1	18.00	0.461	804	89.4	7.07	20.8	6.94	1.14	54.7
S 15 × 50	14.7	15.00	0.550	486	64.8	5.75	15.7	5.57	1.03	50
× 42.9	12.6	15.00	0.411	447	59.6	5.95	14.4	5.23	1.07	42.9

S 12 × 50	14.7	12.00	0.687	305	50.8	4.55	15.7	5.74	1.03	50
× 40.8	12.0	12.00	0.462	272	45.4	4.77	13.6	5.16	1.06	40.8
S 12 × 35	10.3	12.00	0.428	229	38.2	4.72	9.87	3.89	0.980	35
× 31.8	9.35	12.00	0.350	218	36.4	4.83	9.36	3.74	1.00	31.8
S 10 × 35	10.3	10.00	0.594	147	29.4	3.78	8.36	3.38	0.901	35
× 25.4	7.46	10.00	0.311	124	24.7	4.07	6.79	2.91	0.954	25.4
S 8 × 23	6.77	8.00	0.441	64.9	16.2	3.10	4.31	2.07	0.798	23
× 18.4	5.41	8.00	0.271	57.6	14.4	3.26	3.73	1.86	0.831	18.4
S 7 × 20	5.88	7.00	0.450	42.4	12.1	2.69	3.17	1.64	0.734	20
× 15.3	4.50	7.00	0.252	36.7	10.5	2.86	2.64	1.44	0.766	15.3
S 6 × 17.25	5.07	6.00	0.465	26.3	8.77	2.28	2.31	1.30	0.675	17.25
× 12.5	3.67	6.00	0.232	22.1	7.37	2.45	1.82	1.09	0.705	12.5
S 5 × 14.75	4.34	5.00	0.494	15.2	6.09	1.87	1.67	1.01	0.620	14.75
× 10	2.94	5.00	0.214	12.3	4.92	2.05	1.22	0.809	0.643	10
S 4 × 9.5	2.79	4.00	0.326	6.79	3.39	1.56	0.903	0.646	0.569	9.5
× 7.7	2.26	4.00	0.193	6.08	3.04	1.64	0.764	0.574	0.581	7.7
S 3 × 7.5	2.21	3.00	0.349	2.93	1.95	1.15	0.586	0.468	0.516	7.5
× 5.7	1.67	3.00	0.170	2.52	1.68	1.23	0.455	0.390	0.522	5.7

From *AISC Manual of Steel Construction*, Eighth Edition.

PROPERTIES OF W SHAPES FOR DESIGN PURPOSES (WIDE-FLANGE BEAMS)

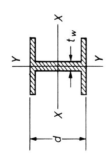

Designation	Area A (in.²)	Depth d (in.)	Web Thickness t_w (in.)	Axis X-X			Axis Y-Y			Nominal Weight per Foot (lb)
				I (in.⁴)	S (in.³)	r (in.)	I (in.⁴)	S (in.³)	r (in.)	
W 10 × 112	32.9	11.36	0.755	716	126	4.66	236	45.3	2.68	112
× 100	29.4	11.10	0.680	623	112	4.60	207	40.0	2.65	100
× 88	25.9	10.84	0.605	534	98.5	4.54	179	34.8	2.63	88
× 77	22.6	10.60	0.530	455	85.9	4.49	154	30.1	2.60	77
× 68	20.0	10.40	0.470	394	75.7	4.44	134	26.4	2.59	68
× 60	17.6	10.22	0.420	341	66.7	4.39	116	23.0	2.57	60
× 54	15.8	10.09	0.370	303	60.0	4.37	103	20.6	2.56	54
× 49	14.4	9.98	0.340	272	54.6	4.35	93.4	18.7	2.54	49
W 10 × 45	13.3	10.10	0.350	248	49.1	4.32	53.4	13.3	2.01	45
× 39	11.5	9.92	0.315	209	42.1	4.27	45.0	11.3	1.98	39
× 33	9.71	9.73	0.290	170	35.0	4.19	36.6	9.20	1.94	33
W 10 × 30	8.84	10.47	0.300	170	32.4	4.38	16.7	5.75	1.37	30
× 26	7.61	10.33	0.260	144	27.9	4.35	14.1	4.89	1.36	26

Handbook Data 417

Shape									
W 10 × 22	6.49	10.17	0.240	118	23.2	4.27	11.4	3.97	1.33	22
× 19	5.62	10.24	0.250	96.3	18.8	4.14	4.29	2.14	0.874	19
× 17	4.99	10.11	0.240	81.9	16.2	4.05	3.56	1.78	0.844	17
× 15	4.41	9.99	0.230	68.9	13.8	3.95	2.89	1.45	0.810	15
× 12	3.54	9.87	0.190	53.8	10.9	3.90	2.18	1.10	0.785	12
W 8 × 67	19.7	9.00	0.570	272	60.4	3.72	88.6	21.4	2.12	67
× 58	17.1	8.75	0.510	228	52.0	3.65	75.1	18.3	2.10	58
× 48	14.1	8.50	0.400	184	43.3	3.61	60.9	15.0	2.08	48
× 40	11.7	8.25	0.360	146	35.5	3.53	49.1	12.2	2.04	40
× 35	10.3	8.12	0.310	127	31.2	3.51	42.6	10.6	2.03	35
× 31	9.13	8.00	0.285	110	27.5	3.47	37.1	9.27	2.02	31
W 8 × 28	8.25	8.06	0.285	98.0	24.3	3.45	21.7	6.63	1.62	28
× 24	7.08	7.93	0.245	82.8	20.9	3.42	18.3	5.63	1.61	24
W 8 × 21	6.16	8.28	0.250	75.3	18.2	3.49	9.77	3.71	1.26	21
× 18	5.26	8.14	0.230	61.9	15.2	3.43	7.97	3.04	1.23	18
W 8 × 15	4.44	8.11	0.245	48.0	11.8	3.29	3.41	1.70	0.876	15
× 13	3.84	7.99	0.230	39.6	9.91	3.21	2.73	1.37	0.843	13
× 10	2.96	7.89	0.170	30.8	7.81	3.22	2.09	1.06	0.841	10
W 6 × 25	7.34	6.38	0.320	53.4	16.7	2.70	17.1	5.61	1.52	25
× 20	5.87	6.20	0.260	41.4	13.4	2.66	13.3	4.41	1.50	20
× 15	4.43	5.99	0.230	29.1	9.72	2.56	9.32	3.11	1.46	15
W 6 × 16	4.74	6.28	0.260	32.1	10.2	2.60	4.43	2.20	0.966	16
× 12	3.55	6.03	0.230	22.1	7.31	2.49	2.99	1.50	0.918	12
× 9	2.68	5.90	0.170	16.4	5.56	2.47	2.19	1.11	0.905	9
W 5 × 19	5.54	5.15	0.270	26.2	10.2	2.17	9.13	3.63	1.28	19
× 16	4.68	5.01	0.240	21.3	8.51	2.13	7.51	3.00	1.27	16
W 4 × 13	3.83	4.16	0.280	11.3	5.46	1.72	3.86	1.90	1.00	13

From AISC Manual of Steel Construction, Eighth Edition.

PROPERTIES OF C SHAPES FOR DESIGN PURPOSES (AMERICAN STANDARD CHANNEL)

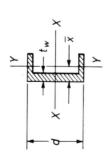

Designation	Area A (in.^{2})	Depth d (in.)	Web Thickness t_w (in.)	Axis X-X I (in.^{4})	S (in.^{3})	r (in.)	Axis Y-Y I (in.^{4})	S (in.^{3})	r (in.)	$\bar{x}$ (in.)	Nominal Weight per Foot (lb)
C 15 × 50	14.7	15.00	0.716	404	53.8	5.24	11.0	3.78	0.867	0.798	50
× 40	11.8	15.00	0.520	349	46.5	5.44	9.23	3.36	0.886	0.777	40
× 33.9	9.96	15.00	0.400	315	42.0	5.62	8.13	3.11	0.904	0.787	33.9
C 12 × 30	8.82	12.00	0.510	162	27.0	4.29	5.14	2.06	0.763	0.674	30
× 25	7.35	12.00	0.387	144	24.1	4.43	4.47	1.88	0.780	0.674	25
× 20.7	6.09	12.00	0.282	129	21.5	4.61	3.88	1.73	0.799	0.698	20.7
C 10 × 30	8.82	10.00	0.673	103	20.7	3.42	3.94	1.65	0.669	0.649	30
× 25	7.35	10.00	0.526	91.2	18.2	3.52	3.36	1.48	0.676	0.617	25
× 20	5.88	10.00	0.379	78.9	15.8	3.66	2.81	1.32	0.692	0.606	20
× 15.3	4.49	10.00	0.240	67.4	13.5	3.87	2.28	1.16	0.713	0.634	15.3
C 9 × 20	5.88	9.00	0.448	60.9	13.5	3.22	2.42	1.17	0.642	0.583	20
× 15	4.41	9.00	0.285	51.0	11.3	3.40	1.93	1.01	0.661	0.586	15
× 13.4	3.94	9.00	0.233	47.9	10.6	3.48	1.76	0.962	0.669	0.601	13.4

Designation											
C 8 × 18.75	5.51	8.00	0.487	44.0	11.0	2.82	1.98	1.01	0.599	0.565	18.75
× 13.75	4.04	8.00	0.303	36.1	9.03	2.99	1.53	0.854	0.615	0.553	13.75
× 11.5	3.38	8.00	0.220	32.6	8.14	3.11	1.32	0.781	0.625	0.571	11.5
C 7 × 14.75	4.33	7.00	0.419	27.2	7.78	2.51	1.38	0.779	0.564	0.532	14.75
× 12.25	3.60	7.00	0.314	24.2	6.93	2.60	1.17	0.703	0.571	0.525	12.25
× 9.8	2.87	7.00	0.210	21.3	6.08	2.72	0.968	0.625	0.581	0.540	9.8
C 6 × 13	3.83	6.00	0.437	17.4	5.80	2.13	1.05	0.642	0.525	0.514	13
× 10.5	3.09	6.00	0.314	15.2	5.06	2.22	0.866	0.564	0.529	0.499	10.5
× 8.2	2.40	6.00	0.200	13.1	4.38	2.34	0.693	0.492	0.537	0.511	8.2
C 5 × 9	2.64	5.00	0.325	8.90	3.56	1.83	0.632	0.450	0.489	0.478	9
× 6.7	1.97	5.00	0.190	7.49	3.00	1.95	0.479	0.378	0.493	0.484	6.7
C 4 × 7.25	2.13	4.00	0.321	4.59	2.29	1.47	0.433	0.343	0.450	0.459	7.25
× 5.4	1.59	4.00	0.184	3.85	1.93	1.56	0.319	0.283	0.449	0.457	5.4
C 3 × 6	1.76	3.00	0.356	2.07	1.38	1.08	0.305	0.268	0.416	0.455	6
× 5	1.47	3.00	0.258	1.85	1.24	1.12	0.247	0.233	0.410	0.438	5
× 4.1	1.21	3.00	0.170	1.66	1.10	1.17	0.197	0.202	0.404	0.436	4.1

From AISC *Manual of Steel Construction*, Eighth Edition.

Answers to Odd-Numbered Problems

Chapter 2

1. 450 lb at section AA, 300 lb at section BB, 150 lb at section CC, 0 lb at section DD
3. $SF = 253$ lb at 191°
5. $X = +206$ lb, $Y = +245$ lb
7. $SF = 523$ lb, $X = +429$ lb
9. System (b) is in equilibrium.
11. $F_1 = 25.9$ lb, $F_2 = 36.6$ lb
13. 216 lb
15. Boom = 1000 lb, wire = 1410 lb
17. $a = 78$ lb, $b = 110$ lb, $c = 95.3$ lb, $d = 55$ lb
19. $F_1 = 36.2$ lb, $F_2 = 51.2$ lb
21. $A = 99$ lb at 53.2°
23. $F_1 = 508$ N, $F_2 = 718$ N

Chapter 3

1. 50 lb
3. $F = 2000$ lb up, couple = 12 000 lb·ft CCW
5. $R_A = 407$ lb up, $R_C = 648$ lb up
7. $R_A = 4475$ lb up, $R_B = 1975$ lb up
9. $R_A = 670$ kN up, $R_B = 1.33$ MN up
11. Torque at $C = 40$ lb·ft CCW
13. $W = 86.7$ lb in (a), $W = 65$ lb in (b)
15. Torque at $C = 9.8$ kN·m CW

17. $R_A = 108$ lb up, $R_B = 45$ lb at $50°$
19. $R_A = 200$ lb up at $67.9°$
21. Force of wire $= 462$ lb, $R_A = 306$ lb down at $319°$
23. $F = 5000$ lb up, couple $= 12\,500$ lb·in. CCW
25. $F = 35.7$ lb down at $303.5°$, couple $= 1086$ lb·in. CCW
27. $X = 4$ ft, $F = 200$ lb

Chapter 4

1. P_1 and P_2
3. $F = 2330$ lb
5. Boom $= 140$ lb C, wire $= 121$ lb T
7. $AB = BC = 490$ lb C, $AC = 315$ lb T
9. Left support $= L$: member $CL = CD = 1.73$ kips C, $AL = 0.87$ kip T, $AC = AB = 1.73$ kips T, $AD = 0$
11. Forces are in kips: (a) $AG = BD = BE = BG = EG = 0$, $AB = BC = 5$ T, $DE = CD = EF = 4$ C; (b) $AB = 2$ T, $AD = BC = 3.16$ T, $BD = 3$ C, $CD = 2$ C, $DE = 4$ C
13. $A_x = 5.95$ kN T, $AB = 6.87$ kN T, $AC = 3.43$ kN C, $BC = 6.87$ kN C
15. Forces on ABC: $A = 500$ lb up, $B_x = 1040$ lb T, $B_y = 0$, $C_x = 1040$ lb C, $C_y = 0$
17. (a) Reaction at $A = 187$ lb at $78.4°$, reaction at $B = 41.1$ lb at $156°$ (reaction is perpendicular to ladder); (b) reaction at $B = -120$ lb, wire at $A = +120$ lb, $A_y = +200$ lb
19. (a) Forces on ABD: $A_y = +250$ lb, $B_y = 0$, $B_x = +750$ lb, $D_x = -750$ lb, $D_y = -250$ lb; (b) forces on ABD: $A_y = +150$ lb, $B_y = 0$, $B_x = +224$ lb, $D_x = -374$ lb, drum force perpendicular to member $= 212$ lb
21. (a) Left member: left reaction $= +73$ lb, right reaction $= +218$ lb; right member: left reaction $= +291$ lb, right reaction $= +109$ lb
(b) Left member: left reaction $= 2.1$ kips, right reaction $= 4.4$ kips; center member: each reaction $= 1.5$ kips; right member: left reaction $= 6$ kips, right reaction $= 1.5$ kips
23. Angle of friction $= 21.8°$
 (a) $F = 200$ lb
 (b) $F = 300$ lb
 (c) $F = 188$ lb
25. (a) Angle of friction $= 38.7°$
 (b) $F = 2400$ lb
27. (a) 51.8 lb, (b) 70.7 lb
29. 1.43 lb
31. (a) 40 lb
 (b) 80 lb
 (c) 120 lb
33. (a) 861 lb
 (b) 673 lb

Chapter 5

1. Measured from lower left corner: $\overline{X} = 3.34$ in., $\overline{Y} = 2.89$ in.
3. Measured from lower left corner: $\overline{X} = \overline{Y} = 3$ in.
5. Centroidal Y axis can be determined by inspection. Centroidal X axis is 4.31 in. above base.
7. Centroidal Y axis can be determined by inspection. Centroidal X axis is 2.25 in. above center of lowest rivet.
9. Total weight = 75.9 lb, cg = 16.2 ft to right of O
11. $I_x = 60.7$ in.^{4}
13. $r_x = 2.60$ in.
15. $r_x = 1.25$ in.

Chapter 6

1. (a) 21 900 psi
   (b) 10 900 psi
   (c) compression
   (d) 17.5 in.^{2}
3. 2700 psi
5. (a) 35 300 lb
   (b) Yes
   (c) 0.389 in.
   (d) Can be made no smaller than 0.643 inch in diameter
7. (a) 2.18 in. diameter
   (b) 2.21 in.
   (c) 75 in.
9. 0.0125
11. (a) 32 000 psi
    (b) 14 400 psi
13. For steel, 37 500 psi; for brass, 20 000 psi; for aluminum, 12 500 psi
15. (a) 0.0193
    (b) 386 mm

Chapter 7

1. (a) Refer to Appendix.
   (b) Refer to Appendix.
3. 43 percent
5. The curve is a sloping straight line to the elastic limit with coordinates of 80 000 psi stress and 0.004 strain. From the elastic limit, the curve is a horizontal line until a strain value of 0.010 is reached.
7. For spring, 0.5 percent carbon steel; for crane hook, low-carbon steel

Answers to Odd-Numbered Problems   423

Chapter 8

1. 6.1 in.²
3. 64 000 lb
5. Square pin must have dimensions of 1.4 in. × 1.4 in. Main link must have dimensions of 3 in. × 1.8 in. Side links must have dimensions of 3 in. × 0.9 in.
7. 2920 lb
9. 9100 lb
11. On cross section, 7500 psi tensile and 0 psi shear; at 9°, 7320 psi tensile and 1160 psi shear; at 18°, 6780 psi tensile and 2200 psi shear; at 27°, 5950 psi tensile and 3050 psi shear; at 36°, 4910 psi tensile and 3570 psi shear; at 45°, 3750 psi tensile and 3750 psi shear
13. 5900 lb

Chapter 9

1. $\delta = \dfrac{\sigma}{E} \cdot L = \dfrac{F/A}{E} \cdot L = \dfrac{F \cdot L}{A \cdot E}$
3. 590 000 lb
5. (a) For copper, 3.9966 in.; for steel, 3.9982 in.
   (b) 1900 lb
7. Three-fourths
9. Brass block = 34 400 lb, each steel block = 32 800 lb
11. (a) 13 300 psi
    (b) For each material, strain = $4.44(10)^{-4}$
    (c) 799 000 lb
13. (a) 455°F
    (b) 262°F
15. (a) 166°F
    (b) 217°F
    (c) 160° and 219°F
17. (a) 26 300 lb
    (b) 5.0007 in.
    (c) 4.9993 in.
19. Pointed tip of rod will be 10.0005 in. above base.
21. (a) 252°F
    (b) 32 000 psi
23. 51.8 MPa

Chapter 10

1. (a) Required bolt area = 0.424 in.²; choose a 1 in. bolt.
   (b) 12.6 bolts needed; choose 13.
3. 425 psi
5. 3500 psi

7. 0.089 in.
9. 0.347 in.
11. (a) No change, no change
    (b) Twice original, twice original
    (c) Twice original, twice original
    (d) Longitudinal seam formula does not apply. Circumferential seam stress is doubled.
13. 320 kPa

Chapter 11

1. (a) 1.21 in.
   (b) 1.51 in.
3. 1270 lb, 2550 lb, 3820 lb, 5100 lb
5. (a) 2.36 in.
   (b) Left side length = 1.89 in.
7. Solution (a), 22.4 in.; solution (b), B = 14.8 in. and C = 5.6 in.; solution (c), 14.8 in.
9. (a) Yes
   (b) 50.4 kN
11. (a) Tensile stress = 25 000 psi, shear stress = 7960 psi, compressive stress = 25 000 psi
    (b) 22 000 lb
13. (a) 21 900 lb
    (b) 66.4 percent
15. (a) 67 900 lb
    (b) 61.7 percent
17. (a) 44 000 lb
    (b) 80 percent
    (c) 293 psi
    (d) 330 psi
19. 90 500 lb
21. Answer varies with loads and torques assumed.
23. Allowable load = 90 kN, efficiency = 40 percent

Chapter 12

1. (a) 6000 psi
   (b) 4500 psi
   (c) Zero
3. 30 MPa
5. 11 400 psi
7. 1.5 in.
9. 2.90 in.
11. 265 MPa
13. 0.934°

15. $\theta = \dfrac{(T \cdot c/J)L}{G \cdot c} = \dfrac{T \cdot L}{J \cdot G}$
17. (a) 2.61°
    (b) Small diameter shaft = 10 200 psi, large diameter shaft = 1270 psi
19. (a) 94 500 lb·in.
    (b) Shear stress = 10 700 psi on shank diameter of bolt, shear stress = 15 600 psi on minor diameter
    (c) Torque = 47 200 lb·in., shear stress = 5350 psi on shank diameter of bolt
21. 3710 N·m

Chapter 13

1. $V_{wall} = -500$ lb
3. $V_{0-5} = -100$ lb, $V_{wall} = -450$ lb
5. Reverse beam so wall is on the right: $V_{0-4} = -100$ lb, $V_{4-7} = -400$ lb, $V_{7-10} = -900$ lb
7. $V_{0-4} = 200$ lb, $V_{4-8} = -800$ lb, $V_{8-12} = 600$ lb
9. $V_{0-8} = -150$ lb, $V_{8-14} = 200$ lb
11. $V_{max} = 1600$ lb at beam ends
13. $V_{0-8} = 488$ lb, $V_{11-14} = -562$ lb
15. $V_{max} = 2943$ N at beam ends
17. $M_{wall} = -10\,000$ lb·ft
19. $M_7 = -2450$ lb·ft, $M_{wall} = -13\,300$ lb·ft
21. $M_8 = 2800$ lb·ft
23. $M_4 = 1000$ lb·ft
25. $M_{4-12} = -4800$ lb·ft
27. $M_{5.15} = -1250$ lb·ft
29. $M_{13.875} = 47\,500$ lb·ft
31. $M_5 = 7360$ N·m
33. $M_X = -100X^2 + 1600X$
35. From 0–6 ft, $M_X = -200X^2 + 7046X$; from 6–14 ft, $M_X = -200X^2 + 5546X + 9000$; from 14–26 ft, $M_X = -200X^2 + 4046X + 30\,000$
37. From 0–2 m, $M_X = -98X^2$; from 2–8 m, $M_X = -98X^2 + 980X - 1960$; from 8–10 m, $M_X = -98X^2 + 1960X - 9800$
39. Place hinge 9.09 ft to the right of R_1.
41. 2.765 m and 7.235 m from left end of beam
43. $A = 600$ lb up, $B = 367$ lb up, $C = 633$ lb up

Chapter 14

1. (a) 15 400 psi
   (b) 7700 psi
3. (a) 24 200 psi T
   (b) 35 100 psi C
   (c) Stresses are inverted but top fibres are still in tension.

5. (a) 11 300 psi
   (b) 5650 psi
7. (a) 563 lb
   (b) 323 lb equals 57 percent of load in part (a)
9. 11 000 psi
11. (a) Cross section = 5.74 in. on a side
    (b) Cross section = 2.54 in. on a side
    (c) Wood beam = 183 lb, steel beam = 439 lb
13. 98.67 MPa
15. 11.9 kg

Chapter 15

1. For Problem 1, 1000 lb; for Problem 5, 500 lb; for Problem 6, 700 lb; for Problem 8(a), 1220 lb; and for Problem 8(b), 2600 lb
3. Dimensions listed are distances from top of beam: 1 in. = 142 psi, 2 in. = 263 psi, 4 in. = 438 psi, 7 in. = 537 psi (at centroidal axis)
5. For Problem 1, 955 psi; for Problem 8(a), 459 psi; for Problem 8(b), 1430 psi; for Problem 10, 1470 psi
7. (a) 427 psi
   (b) 283 psi
9. 2940 psi
11. 38.2 pins needed; round to 39 pins spaced 3.7 in. on center
13. Maximum shear stress = $\tau = 3V/2BH$
15. 438 kPa
17. 2.1 MPa on centroidal axis at wall

Chapter 16

1. $S = bh^2/6$
3. (a) 2 × 16
   (b) 4 × 14
   (c) 2 × 10
5. (a) S 18 × 54.7
   (b) W 10 × 77
7. W 10 × 22
9. 2 × 10
11. S 4 × 9.5

Chapter 17

1. 15 in.
3. 0.008 in.
5. 2750 psi, which exceeds the allowable 1900 psi
7. 1380 mm

Answers to Odd-Numbered Problems   427

9. 1.12 lb
11. 1.84 in.
13. (a) 0.306 in.
    (b) 1.03 in.
15. 151 lb
17. 23.4 mm
19. 0.982 in.
21. 0.0445 in.
23. (a) 0.342 in.
    (b) 0.0296 in.
    (c) 0.069 in.
25. (a) 0.323 in.
    (b) 1.09 in.

Chapter 18

1. Support = 6.25 lb, $R_{wall} = +13.75$ lb, $M_{wall} = 112.5$ lb·in. CW
3. Support = 1556 lb, $R_{wall} = +1444$ lb, $M_{wall} = 4000$ lb·ft CW
5. Support = 90 lb, $R_{wall} = -30$ lb, $M_{wall} = 300$ lb·in. CCW
7. Support = 1230 N, $R_{wall} = 2700$ N, $M_{wall} = 2200$ N·m CW
9. $B = 2750$ lb, $A = C = 625$ lb
11. $B = (11/8) \times P$
13. See Sample Problem 18.6.
15. Each wall supplies a 350 lb force and a 5250 lb·in. moment.
17. Each wall supplies a 700 lb force and an 8820 lb·in. moment.
19. Each wall supplies a 375 lb force and a 312 lb·ft moment.
21. Each wall supplies a 4900 N force and a 3270 N·m moment.

Chapter 19

1. (a) 120
   (b) 98.3
   (c) 265
   (d) 112
3. For part (a), 594 lb; for part (b), 59 000 lb; for part (c), 314 lb
5. (a) 51 800 lb
   (b) 6960 lb
   (c) 1740 lb
7. 9530 lb
9. Yes (allowable load = 514 lb)
11. 328 000 lb
13. S 8 × 18.4
15. 200
17. Critical load = 13 100 N, allowable load = 6550 N

Chapter 20

1. Maximum tensile stress = 5000 psi on a plane 45° CW from AA; maximum compressive stress = 1000 psi on a plane 45° CCW from AA, maximum shear stresses = 3000 psi on plane AA and plane perpendicular to AA
3. Maximum tensile stress on plane AA, no compressive stresses, maximum shear stresses = 5000 psi on planes 45° from AA
5. Maximum tensile stress = 5000 psi on plane AA, maximum compressive stress = 10 000 psi on plane BB, maximum shear stresses = 7500 psi on planes 45° from AA
7. Maximum tensile stress = 6500 psi on a plane 16.8° CW from AA, maximum compressive stress = 11 500 psi on a plane 16.8° CW from BB, maximum shear stresses = 9000 psi on planes 28.2° CCW from AA and BB
9. Maximum tensile stress = 10 000 psi on BB, maximum compressive stress = 6670 psi on AA, maximum shear stresses = 8330 psi on planes 45° from AA
11. Maximum tensile stress = 6370 psi on a plane 45° CW from the cross section plane (plane AA), maximum compressive stress = 6370 psi 45° CCW from AA, maximum shear stresses = 6370 psi on AA and on a plane 90° from AA
13. Maximum tensile stress = 0 psi, maximum compressive stress = 4500 psi
15. Maximum tensile stress = 6000 psi, maximum compressive stress = 12 000 psi
17. Maximum tensile stress = 260 000 psi, maximum compressive stress = 220 000 psi
19. Maximum eccentricity = 1 in.
21. Maximum tensile stress = 22 500 psi, maximum compressive stress = 22 700 psi (These stresses appear on a cross section at base of davit.)
23. 870 lb
25. (a) Shear stress = 54 700 psi
    (b) Shear stress = 37 400 psi
    (c) Shear stress = 174 000 psi
27. Shear stress = 91 300 psi
29. Shear stress = 924 MPa
31. Shear stresses on rivet 1 = 1280 psi; on rivet 4 = 3830 psi
33. Shear stresses on rivet 1 = 1450 psi; on rivet 2 = 276 psi; on rivet 4 = 3730 psi
35. Shear stresses on rivet 1 = 5050 psi; on rivet 2 = 7800 psi; on rivet 3 = 9800 psi
37. Shear stresses on rivet 1 = 990 kPa; on rivet 2 = 2.97 MPa

Index

AISC (American Institute of Steel Construction), 151, 161, 351, 356
AISI (American Iron and Steel Institute), 161
Allowable stress, 150–52, 160, 161–62; in columns, 352–55, 365; for lumber, 407; for steel and aluminum, 150–52, 407
Aluminum: allowable stresses for, 407; mechanical properties of, 137, 138
Aluminum beams, 284
American Institute of Steel Construction (AISC), 151, 161, 351, 356
American Iron and Steel Institute (AISI), 161
American Society of Civil Engineers (ASCE), 161
American Society of Mechanical Engineers (ASME), 161, 404
American Society for Testing and Materials (ASTM), 131, 161
American Welding Society, 199, 201
Analysis, defined, 2
Angle of friction, 79
Angle of twist, 225–28, 232, 234–35
Applied forces, 36
Arc welding, 198
ASCE (American Society of Civil Engineers), 161
ASME (American Society of Mechanical Engineers), 161, 404
ASTM (American Society for Testing and Materials), 131, 161
Axial loads, 167–68, 349

Beam curvature, 302–05, 321–22; formulas for, 304, 320; theory of, 379–80
Beam deflection, 306; of cantilever beams, 312–15, 319–20; computer program for, 325; first moment-area formula for, 306–09, 321; of overhanging beam with end loads, 315–17; problem solving for, 311–12, 322–24; second moment-area formula for, 309–11, 321; of simple beam with eccentric load, 317–19
Beam design, 293–301; beam deflection in, 306; complete shear and moment diagrams in, 343–44, 347; computer program for, 300–01; use of section modulus in, 294–301
Beams: aluminum, 284; bending stresses in, 261–75; built-up, 285–88; cantilever (see Cantilever beams); computer program for, 259–60; defined, 238; elastic curve of, 252–53, 302–03; equilibrium in, 238–40; fixed (see Fixed beams); hinged, 253–54, 256; moment diagrams for, 246–52, 255–56; neutral axis of, 262, 302; overhanging, 244–46, 251–52, 315–17; problem solving with, 238–41; shear diagrams for, 241–46, 247–48, 255; shear stress in, 276–92; Standard I, 414–15; statically determinate, 238, 239; statically indeterminate (see Statically indeterminate beams); steel (see Steel beams); symmetrically loaded, 334, 339, 341–43; two-span continuous, 334–39, 346, 347; wide-flange, 416–17; wooden, 298
Bending stresses, effects of, 261–62
Bending stress formula, 265, 271; applications of, 266–70, 271–74; computer program for, 274–75; conditions for, 265–66; derivation of, 263–65; moments of inertia with, 270–71
Boilers. *See* Pressure vessels.
Bolted covers, of pressure vessels, 184–86, 193–94
Bolted joints, 204–12, 214–16
Bolts (*see also* Pins): dimensions of, 406; in free body diagrams, 17; shear stresses in, 285–88; torque in, 228–30
Brick compression blocks, 378–79
Bridge building, 60
Brinell system, 138
Brittleness, 137
Built-up beams, 285–88
Butt joints: riveted, 204, 209–11; welded, 198–200

Cantilever beams: deflection of, 312–15, 319–20; free body diagram of, 16, 17;

429

moment of force with, 35; parallel force systems with, 36–40; shear and moment diagrams of, 238, 242–44; statically indeterminate, 328
Cantilever truss, 67–72
Carbon steels, 133–34
Cast iron, 136, 137, 151
Center of gravity, 100–03
Centroids, 100–06, 110
Circular shafts: computer program for, 236–37; longitudinal shear force in, 230–31; polar moment of inertia for, 219, 223, 233; shear stress on, 219–20, 233; torsion formulas for, 221–28
Circumferential seams, 186–90, 193–94
Coefficient of friction, 78–79
Coefficent of thermal expansion, 172–73, 406
Coldworking, 140–41
Collinear forces, 11
Column failure, 349, 350–57
Columns: average stress on, 352–55, 365; choice of, 361; classes of, 349–50; with concentric loads, 348–65; defined, 349; end conditions of, 352, 363; load of, 348–49, 360; long, 350–54, 361–62; problem solving procedures for, 357–58; safety factor for, 357; short, 350, 354–56; slenderness ratio for, 349, 359, 362; wood, 356–57
Component, of force, 21–22
Component summation method, 20–25
Compression blocks, 349–50, 362; brick or concrete, 378–79; eccentric loads on, 375–79, 398–400
Compressive forces, 11
Compressive stress, 114–18, 123, 371–74
Concentrated force, 10
Concentrated load, 10
Concentrated stress, 152–55, 161, 163
Concentric load, 350
Concrete, 136, 378
Concurrent force systems, 11, 18–32; component summation method for, 20–25; force polygon method for, 19–20; force triangle method for, 25–27; pulleys as, 40, 41–42
Coplanar forces, 11
Copper, 135, 137
Cosine, 5
Cosine law, 6
Coulomb, Charles, 261
Couples, 13; in parallel force systems, 36; replacement of force system with, 46–47
Couplings, 228–30, 235–36
Cover plates, 209
Creep, 142–44
Critical area, 116, 117–18
Critical load, 350
Critical stress, 350
C shapes, properties of, 418–19
Culmann, Karl, 156, 366
Curved beam theory, 379–80
Curved machine members, 379–83, 398–400

Deflection. *See* Beam deflection.
Deformation, 118–23; due to temperature, 142–43, 172–77, 178–82; and fatigue, 141–42; in statically indeterminate problems, 166–67
Design, defined, 2
Design stress. *See* Allowable stress.
Direct force, 36
Distributed load, 10
Ductility, 137
Dynamics, 10

Eccentric loads: beam deflection with, 317–19; on compression blocks, 375–79, 398–400; formula for, 377, 395; on offset and curved machine members, 379–83, 398–400; on riveted joints, 387–96, 401–02
Elastic curve, 252–53, 302–03
Elasticity, modulus of, 120–21, 133, 135, 409–10
Elastic limit, 135
Endurance limit, 141
Engineering, 1–2
Equation, defined, 3
Equilibrium: in beams, 238–40; concurrent force system in, 19; defined, 10; mechanical, 17–18; nonconcurrent forces in, 44–45; pulleys in, 42–43; statics equations for, 50; of truss joint, 98–99; of two-force member, 61–62
Euler, Leonhard, 350
Euler's formula, 352, 353–54, 361–62
Extensometer, 131, 148

Factor of safety, 151; for columns, 357
Fatigue, 141–42
FBDs (free body diagrams), 14–17
Fillet weld, 200–04

Fink truss, 64
First moment of area, 100–06
Fixed beams, 326–27; at both ends, 339–43, 346; at one end, supported at other, 328–34, 345–46
Force: applied, 36; components of, 21–22; compressive, 11; concentrated, 10; coplanar, 11; defined, 10; direct, 36; moment of, 11–13, 33–36
Force polygon, 19–22
Force systems: concurrent (*see* Concurrent force systems); nonconcurrent, 13, 43–45, 47–49; parallel, 13, 36–40, 41; replacement of, 45–50
Force triangle, 20, 25–27
Force units, 8
Forged axles, 231
Formula, defined, 3
Fractions, working with, 3–5
Framed structures (*see also* Truss structures): defined, 60; problem solving with, 73–76, 92–96, 98; vs. trusses, 73
Free body diagrams (FBDs), 14–17
Friction: angle of, 79; causes of, 77–78; coefficient of, 78–79; formula for, 78; kinetic, 76–77; problem solving with, 79–84, 96–97; static, 76–84; in wedge problems, 84–89

Galileo, 261
Gas welding, 198
Generator shaft, 225
Geometric shapes, properties of, 411–12
Glass, 137
Gold, 135, 137
Graphical calculus, 139
Gravity, center of, 100–03
Great circle, 190
Gyration radius, 108–09, 112, 349, 362

Handbook of Metal Properties (ASME), 409–10
Handbooks: on columns, 349, 354; data from, 404–19; on stress, 161; use of, 1
Hardness, 138
Helical springs, 383–87, 400–01
Hinged beam, 253–54, 256
Hooke, Robert, 118, 120
Hooke's Law, 120, 121, 131, 134
Horsepower, related to torque, 229, 232
Howe truss, 63

Inertia, moment of. *See* Moment of inertia,

Inflection point, 252–53, 256
Internal pressure, 184–86
International System of Units (SI system), 8, 404–05

Joints: butt, 198–200, 204, 209–11; lap, 200–04, 207–09, 212; method of, 64–66; riveted (*see* Riveted joints); welded, 186–90, 198–204, 213–14

Kilogram (unit), 8
Kinetic friction, 76–77
Knife edge, in free body diagrams, 15

Lap joints: riveted, 204, 207–09, 212; welded, 200–04
Lead, 137, 142
Length, and temperature change, 176
Lid, bolts for, 184–86
Line-of-action, 11
Load: axial, 167–68, 349; of columns, 348–49, 360; concentrated, 10; concentric, 350; critical, 350; distributed, 10; eccentric (*see* Eccentric loads); symmetrical, 334, 339, 341–43
Longitudinal seams, 190–93, 194–96
Longitudinal shear force, 230–31
Low-temperature impact, 144
Lug nuts, 34–35
Lumber (*see also* Wood): allowable stresses for, 407; sizes for, 408

Machine members: curved and offset, 379–83, 398–400; fatigue of, 141–42; statically indeterminate, 165–83
Malleability, 137
Manual of Steel Construction (AISC), 407
Mass units, 8
Materials, properties of. *See* Mechanical properties of materials.
Mathematical errors, common, 3–5
Mechanical equilibrium, 17–18
Mechanical properties of materials, 129–48, 409–10; brittleness, 137; in cold-working, 140–41; creep, 142–44; ductility, 137; elastic limit, 135; fatigue, 141–42; hardness, 138; low-temperature impact, 144; malleability, 137; modulus of elasticity, 120–21, 133, 135, 409–10; modulus of resilience, 139; modulus of rigidity, 219, 220, 409–10; modulus of toughness, 139–40; necking range, 136; percent elongation, 409–10;

plastic range, 136; Poisson's ratio, 144–46; proportional limit, 134, 135; static tensile test of, 130–33, 147–48; stiffness, 137; true stress, 138; ultimate tensile strength, 136, 409–10; yield point, 135–36; yield strength, 138, 409–10
Mechanics, defined, 9
Mechanics of Materials (E. P. Popov), 379
Metal fatigue, 141–42
Method of joints, 64–66
Metric system, 2, 8, 404–05
Modulus of elasticity, 120–21, 133, 135, 409–10
Modulus of resilience, 139
Modulus of rigidity, 219, 220, 409–10
Modulus of toughness, 139–40
Mohr, Otto, 157, 366
Mohr's Circle: computer program for, 164, 402–03; construction of, 157–60, 368–69; for stress determination, 366–74, 395, 396–98
Mohs scale, 138
Moment-area formulas, 306–11, 321
Moment arm, 34
Moment center, 11
Moment diagrams, 246–52, 255–56; in beam design, 343–44, 347; in moment-area formulas, 306–11, 321; by parts, 331–34
Moment of area: first, 100–06; second, 106–09
Moment of force: of couples, 13; defined, 11; of nonconcurrent force system, 33–36; vs. torque, 11–13
Moment of inertia, 106–07, 108–09, 110–12; with bending stress formula, 270–71; polar, 107, 219, 223, 233
Motion, Newton's First Law of, 18

Necking range, 136
Neutral axis, 262, 302
New American Machinist's Handbook (LeGrande), 406
Newton (unit), 8
Newton's First Law of Motion, 18
Nickel, 144
Noncurrent force systems, 13, 43–45, 47–49

Offset machine members, 379–83, 398–400
Offset method, 138
Overhanging beams, 244–46, 251–52, 315–17

Parabolas, 413
Parallel force systems, 13, 36–40, 41
Parallelogram Law, 21
Permanent set, 141
Photoelastic stress analysis, 153
Pins (*see also* Bolts): in free body diagrams, 17; in trusses, 63
Plane surface, in free body diagrams, 15–17
Plasticity, 137
Plastic range, 136
Point of inflection, 252–53, 256
Poisson, Siméon, 144
Poisson's ratio, 144–46
Polar moment of inertia, 107, 219, 223, 233
Pound (unit), 8
Pratt truss, 63–64
Pressure vessels: bolted covers of, 184–86, 193–94; circumferential seam of, 186–90, 193–94; computer program for, 196–97; longitudinal seam of, 190–93, 194–96
Problem solving procedures, 7–8; with beam deflection, 311–12; with beam fixed at one end, supported at other, 328–29; with beams, 238–41; with centroids, 104–06; with columns, 357–58; with framed structures, 73–76; with friction, 79–84; with nonconcurrent force systems, 43–45; with riveted butt joints, 209–10; with riveted lap joints, 207–08; with statically indeterminate problems, 167; with stress/strain problems, 121–22; with temperature problems, 173; with truss structures, 64–66; with welded joints, 202
Projected plane area, 190
Proportional limit, 134, 135
Pulleys, 40–43

Radian, 6
Radius of gyration, 108–09, 112, 349, 362
Reaction, defined, 10
Resilience, modulus of, 139
Rigid body, 10
Rigidity, modulus of, 219, 220, 409–10
Riveted joints, 204–12, 214–17; eccentrically loaded, 387–96, 401–02
Rockwell system, 138
Roller, in free body diagrams, 15
Roof framing, 98
Rounding off, 6–7

Safety factor, 151; for columns, 357

INDEX 433

Saw blade, 305
Scalar quantity, 10
Seams, of pressure vessels: circumferential, 186–90, 193–94; longitudinal, 190–93, 194–96
Second moment of area, 106–09
Section modulus, 294, 298
Shear diagram, 241–46, 255; in beam design, 343–44, 347; relationship to moment diagram, 247–48
Shearing modulus. *See* Modulus of rigidity.
Shear stress, 114–18, 123; angle of twist related to, 225–28, 232; in beams, 276–92, in bolts, 285–88; on circular shafts, 219–20, 233; formula for, 116, 123, 161; longitudinal, 230–31; Mohr's Circle to determine, 370–73; related to torque, 221–25, 232, 233–34; in round wire helical springs, 383–87
Sign convention, 13–14
Significant figures, 6–7
Silver, 137
Simpson's Rule, 139
Sine, 5
Sine law, 6
SI system, 8, 404–05
Slenderness ratio, 349, 359, 362. *See also* Columns.
Slug (unit), 8
Spring formulas, 383–85, 395
Spring index, 385–86
Springs, helical, 383–87, 400–01
S shapes, properties of, 414–15
Standard Handbook for Mechanical Engineers (Baumeister & Marks), 406, 407, 409–10
Statically determinate beams, 238, 239
Statically indeterminate beams, 238, 239, 326; fixed at both ends, 339–43, 346; fixed at one end, supported at other, 328–34, 345–46; two-span continuous, 334–39, 346, 347
Statically indeterminate machine members, 165–66; with axial loads, 167–68, 177–78; computer program for, 183; effect of temperature change on, 172–76, 178–82; problem solving with, 167, 169–72
Static friction, 76–84
Statics, defined, 2, 10
Static tensile test, 130–33, 147–48
Steel: allowable stress for, 150–52, 407; brittleness of, 137; carbon, 133–34; coldworking of, 140–41; concentrated stress of, 152–55; creep in, 143; deformation in, 143; factor of safety for, 151; low-temperature impact on, 144; modulus of resilience for, 140; proportional limit for, 134; stress-strain curve for, 131; toughness of, 140; types of, 133–34; welded, 200; yield point for, 135
Steel beams: C-shaped, 418–19; design of, 295–98, 300–01; shear stress formula for, 284; specifications for, 150; S-shaped, 295, 296–97, 414–15; W-shaped, 295, 297, 416–17
Stiffness, 137
Strain formula, 119, 122, 123
Strength of materials, 2, 113; in columns, 350, 362
Stress: allowable (*see* Allowable stress); compressive, 114–18, 123, 371–74; concentrated, 152–55, 161, 163; critical, 350; shear (*see* Shear stress); from two-force system, 155–60, 161, 163–64; true, 138
Stress formula, 116, 122, 123
Stress–strain curve, 131
Stress–strain formula, 120–21, 123
Stress–strain relationship: coldworking and, 140–41; computer program for, 147–48; modulus of elasticity in, 133; modulus of resilience in, 139; modulus of toughness in, 139–40; proportional limit, above, 134; in statically indeterminate members, 166, 176; in static tensile test, 131–32, 147–48; true stress and, 138; ultimate strength in, 136; yield point in, 135–36; yield strength in, 138; in Young's formula, 120–21
Structural Handbook (Alcoa), 407
Supports, 326
Symmetrical loads, 334, 339, 341–43

Tangent, 5, 6
Tear out, 207
Temperature, effects of, 142–44, 172–77, 178–82
Tensile force, 11
Tensile strength, ultimate, 136, 409–10
Tensile stress, 114–18, 123; on circumferential seam, 186–90; on longitudinal seam, 190–93; Mohr's Circle to determine, 371–74
Term, defined, 4–5
Thermal expansion, coefficient of, 172–73, 406

Thin-walled pressure vessels. *See* Pressure vessels.
Throat dimension, 200
Torque: angle of twist related to, 226–27, 232; defined, 11–13, 221; horsepower related to, 227, 229, 232; shear stress related to, 221–25, 233–34; symbol for, 218
Torque arm, 34
Torsion, defined, 218
Torsion formulas, 232; for circular shafts, 221–28; relating angle of twist to stress, 225–28, 232; relating angle of twist to torque, 226–27, 232; relating horsepower to torque, 227, 229, 232; relating shear stress to torque, 221–25, 232, 233–34
Toughness, modulus of, 139–40
Trigonometry, 5–6
True stress, 138
Truss structures (*see also* Framed structures): cantilever, 67–72; defined, 62–63; equilibrium of, 98–99; problem solving with, 64–66, 89–92; types of, 63–64
Twist, angle of, 225–28, 232, 234–35
Two-force members, 61–62, 89–92
Two-force system, 155–60, 161, 163–64
Two-span continuous beams, 334–39, 346, 347

Ultimate tensile strength, 136, 409–10
Universal testing machine, 130–31
U.S. Customary System (USCS), 8

Varignon, Pierre, 35
Varignon's Theorem, 35, 36
Vector quantity, 10
Vector summation, 21–22
Vickers system, 138

Wahl, A. M., 385
Wandmacher, Dean C., 404
Wedges, 84–89, 97–98
Welded joints: in cylindrical vessels, 186–90; in metal plates, 198–204, 213–14
Wood (*see also* Lumber): beams, 298; columns, 356–57
Working stress. *See* Allowable stress.
W shapes, properties of , 416–17

Yield point, 135–36
Yield strength, 138, 409–10
Young, Thomas, 120
Young's formula, 120–21

Zero-force members, 66–67